T0122212

Springer Theses

Recognizing Outstanding Ph.D. Research

Aims and Scope

The series "Springer Theses" brings together a selection of the very best Ph.D. theses from around the world and across the physical sciences. Nominated and endorsed by two recognized specialists, each published volume has been selected for its scientific excellence and the high impact of its contents for the pertinent field of research. For greater accessibility to non-specialists, the published versions include an extended introduction, as well as a foreword by the student's supervisor explaining the special relevance of the work for the field. As a whole, the series will provide a valuable resource both for newcomers to the research fields described, and for other scientists seeking detailed background information on special questions. Finally, it provides an accredited documentation of the valuable contributions made by today's younger generation of scientists.

Theses are accepted into the series by invited nomination only and must fulfill all of the following criteria

- They must be written in good English.
- The topic should fall within the confines of Chemistry, Physics, Earth Sciences, Engineering and related interdisciplinary fields such as Materials, Nanoscience, Chemical Engineering, Complex Systems and Biophysics.
- The work reported in the thesis must represent a significant scientific advance.
- If the thesis includes previously published material, permission to reproduce this must be gained from the respective copyright holder.
- They must have been examined and passed during the 12 months prior to nomination.
- Each thesis should include a foreword by the supervisor outlining the significance of its content.
- The theses should have a clearly defined structure including an introduction accessible to scientists not expert in that particular field.

More information about this series at http://www.springer.com/series/8790

Philipp Jörg

Exploring the Size of the Proton

by Means of Deeply Virtual Compton Scattering at CERN

Doctoral Thesis accepted by
the Albert Ludwigs University of Freiburg, Freiburg,
Germany

 Springer

Author
Dr. Philipp Jörg
Faculty for Mathematics and Physics
Albert Ludwigs University of Freiburg
Freiburg
Germany

Supervisor
Prof. Horst Fischer
Faculty for Mathematics and Physics
Albert Ludwigs University of Freiburg
Freiburg
Germany

ISSN 2190-5053   ISSN 2190-5061   (electronic)
Springer Theses
ISBN 978-3-030-07983-3   ISBN 978-3-319-90290-6   (eBook)
https://doi.org/10.1007/978-3-319-90290-6

This Springer imprint is published by the registered company Springer International Publishing AG
part of Springer Nature
The registered company address is: Gewerbestrasse 11, 6330 Cham, Switzerland

Supervisor's Foreword

A PhD project in subatomic physics can be like an excursion. At first, it is nothing more than a vague, sketchy, maybe even crazy and imprecise idea about a groundbreaking experiment to gain insight into a new or longstanding physics question. Then the idea gains some traction, preparations start, organization takes over, but the goal seems too ambitious and out of reach, and briefly takes the wind out of your sails. However, with a fresh puff the initial excitement comes back as you take deliberate actions and head off towards uncharted territories. From here out it is very meticulous work to achieve progress and often you go in the opposite direction. But, you persevere and ultimately arrive at your new discovery; the initial question has found its answer and the final publication is submitted. This is the moment to celebrate; the pinnacle of a student's education. Dr. Jörg covers precisely such a journey with his thesis. It starts out with a question, lays the foundation of the known, and describes the measurement, followed by the analysis. As a reward to the reader, Dr. Jörg shares his first-hand insight of a newly captured picture of the three-dimensional structure of the proton, never seen before.

Understanding the structure of the proton is a fundamental challenge and one of the unsolved mysteries that physics faces today. A typical tool for experimentally accessing the internal structure of the proton is lepton–nucleon scattering. In particular, deeply-virtual Compton scattering at large photon virtuality and small four-momentum transfer to the proton provides an excellent tool to obtain a three-dimensional tomographic picture of the proton. Using clear language, Dr. Jörg presents the highly complex subject of this pioneering measurement taken at CERN in a manner suited for freshmen and experts alike. In detail, he provides the foundations of the measurement, the data analysis, and includes exhaustive studies of potential systematic uncertainties, which could bias the result.

This thesis is a rare jewel, describing fundamental research in a highly dynamic field of subatomic physics. It will serve as a map for future followers to travel similar journeys exploring the structure of the proton and enjoying the beauty of particle physics.

Freiburg, Germany   Prof. Dr. Horst Fischer

Acknowledgements

I would like to thank everybody who contributed to the success of this work:

- Prof. Horst Fischer, for assigning me to this interesting topic and for his comprehensive support.
- Prof. Kay Königsmann, for the hospitality in his working group.
- All members of the COMPASS collaboration, for making this measurement possible. I gratefully acknowledge your skills and efforts. In particular, I would like to thank: Dr. Andrea Ferrero, Dr. Eric Fuchey, Dr. Nicole D'Hose, and Prof. Andrzej Sandacz.
- Dr. Florian Herrmann and Dr. Sebastian Schopferer, who always had an open ear to listen to me.
- Nina Hirschinger and Matthias Gorzellik, for the proofreading of this work.
- All members of our working group, for the pleasant atmosphere. In particular, I would like to thank Matthias Gorzellik for the constructive teamwork and the many interesting discussions we had.
- My parents, for the support during my studies.

Ein ganz besonderer Dank gilt meiner Freundin, Nina Hirschinger, für die kontinuierliche Unterstützung während der nicht immer einfachen Zeit, in der diese Arbeit entstanden ist.

Contents

Chapter 1
Preamble

It is an inherent part of human nature to decompose and understand the objects which surround us. The answer to the question of what matter is made of and how its macroscopic properties can be explained on a microscopic level pushed the technical and theoretical boundaries further and further.

The first discovery of what is regarded with today's knowledge as one of the elementary constituents of matter was achieved by J. J. Thomson in 1897. Within cathode-ray-experiments he could proof the existence of the electron [1]. The atom, which was regarded as the elementary building block of matter so far, started to show a substructure. It should not take more than 12 years and Thomson's picture that the atom is build of electrons, surrounded by a massless positive medium [2], could be ruled out. The observation of large scattering angles within the scattering of α particles of gold atoms lead the working group around E. Rutherford to the discovery of the positively charged nuclear core, consisting of protons [3]. The experimental proof of the existence of the neutron in 1932 by J. Chadwick [4] completed the picture of the atom. But this should not mark the end of the story. Nature did not reveal all its secrets back then and neither does it today.

The deviation of the magnetic moment of the nucleon from a pointlike spin-$\frac{1}{2}$ Dirac particle was the first evidence that the nucleon is not an elementary building block of matter either [5, 6]. A dedicated study, which marked the beginning of the age of particle accelerators, reveiled the size of the nucleon in elastic electron nucleon scattering experiments [7]. The nucleon could clearly not be regarded as pointlike anymore and the technique of elastic scattering was soon extended to inelastic and deep inelastic scattering. This lead to the observation of the so called scaling behaviour of the measured cross sections [8, 9] and to the discovery of a variety of new particles during the following years. The structuring of this variety of particles and the explanation of the scaling behaviour were triggered by M. Gell-Mann, G. Zweig and A. Peterman who postulated that the nucleon is build of fundamental pointlike particles [10–12], which are referred to as partons or quarks, antiquarks and gluons in the modern literature.

© Springer International Publishing AG, part of Springer Nature 2018
P. Jörg, *Exploring the Size of the Proton*, Springer Theses,
https://doi.org/10.1007/978-3-319-90290-6_1

Today it is well established that all visible matter is build of baryons and the three generations of leptons. The baryons are further classified into hadrons and mesons, which are build of quarks, antiquarks and gluons. Similar to the leptons, quarks are structured into three generations. The two quarks and antiquarks of each generation appear in three states of the so called fundamental colour charge and are subject to the strong interaction via the exchange of gluons. With this in mind, the first part of the initial question, what matter is made of, seems to have found an answer and the second part, of how to explain the macroscopic properties of matter on the microscopic level, can be attacked.

Tremendous efforts have been made to understand the Englert-Brout-Higgs-Guralnik-Hagen-Kibble mechanism, which led to the successful discovery of the Higgs Boson [13, 14] and the clarification of the orgin of the mass of fundamental particles. However, it is often forgotten that the vast majority of visible matter is given by baryons, which gain most of their mass dynamically within poorly known non-perturbative Quantum Chromo Dynamics processes. The best laboratory to study the underlying mechanisms of non-perturbative Quantum Chromo Dynamics is still given by the nucleon and the central question of how the macroscopic properties of a nucleon like its mass, spin and size can be comprehensively decomposed into the microscopic description in terms of quarks, antiquarks and gluons remains still open.

Dedicated to the decomposition of the spin of the nucleon several experiments at CERN,[1] DESY[2] and SLAC[3] have been carried out, while a lot of today's interest should still be attributed to the early findings of the EMC[4] collaboration. The EMC collaboration observed that only a small part of the spin of the nucleon is given by the quarks and antiquarks [15]. These findings are in strong contrast to the naive quark parton model, which states that the spin of the nucleon is mainly originating from the spins of its three valence quarks and predicts even in relativistic extensions a contribution from the valence quarks of about 60 percent. Within the framework of inclusive and semi-inclusive deep inelastic scattering the qualitative statement of the EMC was experimentally verified [16, 17] and further decomposed during the following years. As of today, flavour specific contributions of the different quarks to the spin of the nucleon are determined [18–21] and the assumption of a very large contribution of the gluons to the spin of the nucleon is ruled out [22, 23]. Triggered by the findings of the EMC experiment and the recent experimental results, it also became evident that a one dimensional description of the nucleon in terms of parton helicity distributions will never lead to a comprehensive picture of its spin decomposition as for example the concept of orbital angular momenta [24] can not be in-cooperated in a one dimensional description.

Extending beyond the scope of inclusive and semi-inclusive deep inelastic scattering, so called Generalised Parton Distributions can be accessed in exclusive deep

---

[1] Conseil Européen pour la Recherche Nucléaire.

[2] Deutsches Elektron Synchrotron.

[3] Stanford Linear Accelerator Center.

[4] European Muon Collaboration.

inelastic scattering experiments. A major part of the COMPASS-II[5] physics programme is dedicated to the investigation of Generalised Parton Distributions, which aim for the most complete description of the partonic structure of the nucleon, comprising both, spatial and kinematic distributions. By including transverse degrees of freedom a three dimensional picture of baryonic matter is created, which will revolutionise our understanding of what comprises 99 percent of the visible matter. Generalised Parton Distributions are experimentally accessible via lepton-induced exclusive reactions, in particular the Deeply Virtual Compton Scattering (DVCS) and Hard Exclusive Meson Production (HEMP). At COMPASS-II those processes are investigated using a high intensity muon beam with a momentum of 160 GeV/c together with a 2.5 m-long liquid hydrogen target, surrounded by the target time of flight system CAMERA[6] and an open field two stage spectrometer, to detect and identify charged and neutral particles.

After a discussion of theoretical and experimental methods related to the structure of the nucleon within Chap. 2 and a description of the COMPASS-II experiment in Chap. 3 the actual scientifc contribution of this thesis is outlined. It comprises the DVCS analysis of the data recorded in 2012 within the framework of a pilot run for the dedicated 2016/2017 DVCS data taking as well as vital improvements on the CAMERA prototype used in 2012.

Chapter 4 will summarise the application of a kinematically constrained fit to the COMPASS-II data, which provides an essential tool within the whole analysis. Chapter 5 consists of a detailed description of the calibration of the CAMERA detector. Furthermore, the determination of the luminosity, the application of data quality criteria and the determination of the efficiency of the CAMERA detector are described throughout this chapter. An overview of the available Monte Carlo simulation techniques and a detailed description of the selection of the exclusive single photon sample is given in Chap. 6. The analysis concludes with Chap. 7, comprising the extraction of the DVCS cross section and its dependence on the square of the four-momentum transfer to the target proton as well as the treatment of the related systematic uncertainties and an interpretation of the results.

Within the concept of Generalised Parton Distributions the square of the four-momentum transfer to the target proton is closely related to the transverse size of the nucleon. The pioneering measurement carried out within this thesis will give a first evaluation of the transverse size of the nucleon as a function of the Bjorken scaling variable x_{Bj} in the uncharted territory of $10^{-2} < x_{\mathrm{Bj}} < 0.2$.

The exclusive measurement of DVCS demands an efficient and precise detection of the recoiled target nucleon, which is achieved by the CAMERA detector. The extensive detector performance studies, carried out within this thesis, lead to vital improvements on the CAMERA detector prototype used in 2012. The application of these improvements, resulting in the successful detector commissioning during the beginning of the dedicated DVCS measurement, are discussed in Chap. 9.

---

[5]**CO**mmon **M**uon **P**roton **A**pparatus for **S**tructure and **S**pectroscopy.

[6]**COMPASS A**pparatus for **M**easurements of **E**xclusive **ReA**ctions.

References

1. J.J. Thomson, XL. Cathode rays. The London, Edinburgh, and Dublin Philos. Mag. J. Sci. **44**, 293–316 (1897). https://doi.org/10.1080/14786449708621070
2. J.J. Thomson, XXIV. On the structure of the atom: an investigation of the stability and periods of oscillation of a number of corpuscles arranged at equal intervals around the circumference of a circle; with application of the results to the theory of atomic structure. The London, Edinburgh, and Dublin Philosophical Magazine and Journal of Science **7**, 237–265 (1904). https://doi.org/10.1080/14786440409463107
3. E. Rutherford, LXXIX. The scattering of α and β particles by matter and the structure of the atom. The London, Edinburgh, and Dublin Philos. Mag. J. Sci. **21**, 669–688 (1911). https://doi.org/10.1080/14786440508637080
4. J. Chadwick, Possible existence of a neutron. Nature **129**, 312 (1932). https://doi.org/10.1038/129312a0
5. R. Frisch, O. Stern, Über die magnetische Ablenkung von Wasserstoffmolekülen und das magnetische Moment des Protons. I. Zeitschrift für Physik **85**, 4–16 (1933). https://doi.org/10.1007/BF01330773
6. L.W. Alvarez, F. Bloch, A quantitative determination of the neutron moment in absolute nuclear magnetons. Phys. Rev. **57**, 111–122 (1940). https://doi.org/10.1103/PhysRev.57.111
7. R. Hofstadter, Electron scattering and nuclear structure. Rev. Mod. Phys. **28**, 214–254 (1956). https://doi.org/10.1103/RevModPhys.28.214
8. E.D. Bloom et al., High-energy inelastic $e - p$ scattering at $6°$ and $10°$. Phys. Rev. Lett. **23**, 930–934 (1969). https://doi.org/10.1103/PhysRevLett.23.930
9. M. Breidenbach et al., Observed behavior of highly inelastic electron-proton scattering. Phys. Rev. Lett. **23**, 935–939 (1969). https://doi.org/10.1103/PhysRevLett.23.935
10. M. Gell-Mann, A schematic model of Baryons and Mesons. Phys. Rev. Lett. **8**, 214–215 (1964). https://doi.org/10.1016/S0031-9163(64)92001-3
11. G. Zweig, An SU(3) model for strong interaction symmetry and its breaking, in *CERN-TH-401* (1964), http://cds.cern.ch/record/570209/files/CERN-TH-412.pdf
12. A. Petermann, Propriétés de l'étrangeté et une formule de masse pur les mésons vectoriels. Nucl. Phys. **63**, 349–352 (1965). https://doi.org/10.1016/0029-5582(65)90348-2
13. C.M.S. Collaboration, S. Chatrchyan et al., Observation of a new boson at a mass of 125 GeV with the CMS experiment at the LHC. Phys. Lett. B **716**, 30–61 (2012). https://doi.org/10.1016/j.physletb.2012.08.021
14. ATLAS Collaboration, G. Aad et al., Observation of a new particle in the search for the Standard Model Higgs boson with the ATLAS detector at the LHC. Phys. Lett. B **716**, 1–29 (2012). https://doi.org/10.1016/j.physletb.2012.08.020
15. E.M.C. Collaboration, J. Ashman et al., A measurement of the spin asymmetry and determination of the structure function g(1) in deep inelastic Muon-Proton scattering. Phys. Lett. B **206**, 364 (1988). https://doi.org/10.1016/0370-2693(88)91523-7
16. COMPASS Collaboration, C. Adolph et al., The spin structure function g_1^p of the proton and a test of the Bjorken sum rule. Phys. Lett. B **753**, 18–28 (2016). https://doi.org/10.1016/j.physletb.2015.11.064
17. COMPASS Collaboration, C. Adolph et al., Final COMPASS results on the deuteron spin-dependent structure function g_1^d and the Bjorken sum rule (2016), arXiv:1612.00620
18. HERMES Collaboration, A. Airapetian, et al., Quark helicity distributions in the nucleon for up, down and strange quarks from semi-inclusive deep-inelastic scattering. Phys. Rev. D **71**, 012003 (2005). https://doi.org/10.1103/PhysRevD.71.012003
19. HERMES Collaboration, A. Airapetian, et al., Measurement of Parton distributions of strange quarks in the nucleon from charged-Kaon production in deep-inelastic scattering on the Deuteron. Phys. Lett. B **666**, 446–450 (2008). https://doi.org/10.1016/j.physletb.2008.07.090
20. SMC Collaboration, B. Adeva et al., Polarised quark distributions in the nucleon from semi-inclusive spin asymmetries. Phys. Lett. B **420**, 180–190 (1998), ISSN 0370-2693. https://doi.org/10.1016/S0370-2693(97)01546-3

21. COMPASS Collaboration, M.G. Alekseev et al., Quark helicity distributions from longitudinal spin asymmetries in muon–proton and muon–deuteron scattering. Phys. Lett. B **693**, 227–235 (2010). https://doi.org/10.1016/j.physletb.2010.08.034
22. COMPASS Collaboration, C. Adolph et al., Leading-order determination of the gluon polarisation using a novel method (2015), arXiv:1512.05053
23. D. de Florian et al., Evidence for polarization of gluons in the proton. Phys. Rev. Lett. **113**, 012001 (2014). https://doi.org/10.1103/PhysRevLett.113.012001
24. R.L. Jaffe, A. Manohar, The g1 problem: deep inelastic electron scattering and the spin of the proton. Nucl. Phys. B **337**, 509–546 (1990). https://doi.org/10.1016/0550-3213(90)90506-9

Chapter 2
Introduction to Theory

This chapter is supposed to give an overview of the theoretical and experimental knowledge on the structure of the nucleon. After a short introduction to elastic scattering and the topic of Form Factors, inclusive and semi-inclusive deep inelastic scattering techniques are explained. The focus is put on the spin decomposition of the nucleon, which will motivate the subject of Generalised Parton Distributions. The chapter concludes with the introduction of Generalised Parton Distributions and the description of a particular exclusive deep inelastic scattering process, Deeply Virtual Compton scattering, which is the most clean channel to constrain Generalised Parton Distributions experimentally.

2.1 Elastic Scattering and Form Factors

Since the first measurement of the magnetic moment of the proton by O. Stern [1] the hypothesis of the proton being a pointlike particle could be excluded, due to the significant deviation of the experimental result to the magnetic moment of a spin-$\frac{1}{2}$ Dirac particle. M. N. Rosenbluth was the first one, who discussed the possibility that an electron, being elastically scattered of a proton, is influenced by reduced charges and reduced magnetic moments. He connected this to the fact that the proton is build by a neutron core and a positively charged meson cloud [2]. Though his picture of the proton itself did not establish, the idea of reduced effective charges and magnetic moments was carried on by R. Hofstadter. He explained the results of Ref. [3] for the differential ep cross section with the Mott cross section, being modified by a phenomenological Form Factor $F(q)$. This Form Factor is related to the charge distribution $\rho(r)$ of the proton by a Fourier transformation [4]. In modern notation his approach reads [5]:

© Springer International Publishing AG, part of Springer Nature 2018
P. Jörg, *Exploring the Size of the Proton*, Springer Theses,
https://doi.org/10.1007/978-3-319-90290-6_2

$$\frac{d\sigma}{d\Omega} = \left(\frac{\alpha^2(\hbar c)^2}{4E^2 \sin^4\left(\frac{\theta}{2}\right)}\right) \cdot \left(1 - \beta^2 \sin^2\left(\frac{\theta}{2}\right)\right) \cdot |F(q)|^2$$

$$:= \left(\frac{d\sigma}{d\Omega}\right)_{Rutherford} \cdot \left(1 - \beta^2 \sin^2\left(\frac{\theta}{2}\right)\right) \cdot |F(q)|^2$$

$$:= \left(\frac{d\sigma}{d\Omega}\right)_{Mott}^* \cdot |F(q)|^2 = \left(\frac{d\sigma}{d\Omega}\right)_{Mott}^* \left|\int_{volume} \rho(\vec{r})e^{i\vec{q}\vec{r}}d^3\vec{r}\right|,$$

while the incident electron energy is denoted by E, the electromagnetic coupling constant by α, the electron velocity in units of the speed of light c by β, the Planck constant by $\hbar$, the magnitude of the centre of mass momentum transfer between the incident and scattered electron by $q = |\vec{q}|$ and the polar scattering angle of the electron by θ. The "$*$" emphasises the fact that the recoil of the target is not taken into account within this formula. Assuming an exponential distribution for $\rho(r)$, this phenomenological ansatz described the data quite well. However, as the proton has a charge and a magnetic distribution, it is easy to judge with today's knowledge that a single Form Factor can not give a complete description.

It is shown from first principles within the one-photon-exchange approximation that the calculation of the cross section of elastic electron proton scattering can be separated into a leptonic and a hadronic part. A complete description of the latter, satisfying Lorentz invariance, symmetry under space reflection and charge conservation, is given by two real functions [6]. In modern notation the cross section of elastic electron-proton scattering is given by [5]:

$$\frac{d\sigma}{d\Omega} = \left(\frac{d\sigma}{d\Omega}\right)_{Mott}^* \cdot \frac{E'}{E} \cdot \left(G_E^2(Q^2) + \frac{\tau}{\epsilon}G_M^2(Q^2)\right)/(1+\tau)$$

$$:= \left(\frac{d\sigma}{d\Omega}\right)_{Mott} \cdot \left(G_E^2(Q^2) + \frac{\tau}{\epsilon}G_M^2(Q^2)\right)/(1+\tau), \tag{2.1}$$

while Q^2 is the negative of the square of the four-momentum transfer to the scattered electron. Originally the Dirac and Pauli Form Factors F_1 and F_2 were introduced, which are related to the electric and magnetic Sachs Form Factors G_E and G_M [7], used today, by:

$$G_E = F_1 - \tau F_2 \text{ and } G_M = F_1 + F_2.$$

The quantity τ is given by $\tau = \frac{Q^2}{4M^2c^2}$, while M denotes the mass of the proton or respectively the neutron. The virtual photon polarisation ϵ is given by:

$$\epsilon = \left(1 + 2(1+\tau)\tan^2\left(\frac{\theta}{2}\right)\right)^{-1}.$$

From Eq. 2.1 it is possible to disentangle $G_E(Q^2)$ and $G_M(Q^2)$ by building a reduced cross section $\left(\frac{d\sigma}{d\Omega}\right)_r$, using the experimentally measured cross section $\left(\frac{d\sigma}{d\Omega}\right)_{Exp.}$:

$$\left(\frac{d\sigma}{d\Omega}\right)_r = \frac{\epsilon(1+\tau)}{\tau}\left(\frac{d\sigma}{d\Omega}\right)_{Exp.} / \left(\frac{d\sigma}{d\Omega}\right)_{Mott} = G_M^2 + \frac{\epsilon}{\tau}G_E^2.$$

The linear dependence on ϵ is exploited, as for a fixed value of Q^2 the slope and the intercept of the reduced cross section are given by $\frac{1}{\tau}G_E^2$ and respectively G_M^2. This technique is commonly known as the Rosenbluth separation.

2.1.1 The Radius of the Proton

Applying the interpretation of R. Hofstadter to the electric and magnetic Form Factors for sufficiently small values of Q^2, for which $Q^2 \approx \vec{q}^2$, the Form Factors $G_E(Q^2)$ and $G_M(Q^2)$ can be interpreted as the Fourier transforms of the charge and magnetic distributions. The assumption of a charge and magnetic distribution, which decrease exponentially with respect to their centre, leads to the so called standard dipole parametrisation, $G_{std.dipole}$, of the Form Factors:

$$G_E = \frac{G_M}{\mu_p} = G_{std.dipole} = \left(1 + \frac{Q^2}{0.71(\text{GeV/c})^2}\right)^{-2}. \tag{2.2}$$

The magnetic moment of the proton divided by the nuclear magneton is denoted by μ_p. Measuring the precise dependence of $G_{E/M}$ on Q^2 close to zero, the mean electric and magnetic proton radius squared can be extracted as [8]:

$$< r_{E/M}^2 >= -\frac{6\hbar^2}{G_{E/M}(0)} \cdot \frac{dG_{E/M}(Q^2)}{dQ^2}.$$

Figure 2.1 shows a high precision measurement of the electric Form Factor, performed at the Mainz accelerator MAMI,[1] which is compared to recent measurements and fits.

Within Ref. [8] the Form Factors were extracted by a direct least squares fit of a variety of different models to the measured ep cross section data and cross checked within the Rosenbluth separation technique, mentioned in Sect. 2.1. The extracted values of the electric r_E and magnetic r_M radii are given as:

$$\sqrt{< r_E^2 >} = 0.879(5)_{stat}(4)_{sys}(2)_{model}(4)_{group} \text{ fm},$$

$$\sqrt{< r_M^2 >} = 0.777(13)_{stat}(9)_{sys}(5)_{model}(2)_{group} \text{ fm},$$

---

[1]**MA**inzer **MI**krotron.

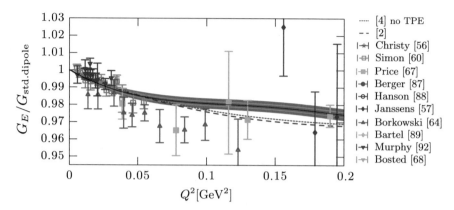

Fig. 2.1 The Form Factor G_E normalised to the standard dipole $G_{std.dip.}$ according to Eq. 2.2 as measured by Ref. [8]. Black line: Best fit to the Mainz data. Blue area: Statistical 68% point wise confidence band. Light blue area: Experimental systematic error. Green outer band: Variation of the Coulomb correction by ±50%. The indicated references are given within Ref. [8]

while the group error refers to a deviation between the two groups of models using spline and polynomial techniques.

The technique of elastic scattering is not the only way to determine the charge radius of the proton though. An alternative approach lies within the measurement of the hyperfine structure and the lamb shift of hydrogen atoms, which has been extended in 2010 by measuring the energy difference between the $2S_{1/2}$ and $2P_{1/2}$ states of muonic hydrogen [9]. Table 2.1 shows a comparison of the electric charge radius of the proton, using different measurement techniques. While the measurements using an electron seem to be compatible between each other, there is an obvious discrepancy with respect to the muonic measurements. This is commonly known as the "Proton Radius Puzzle".

A recent measurement of the 2S-2P transition in muonic deuterium [11] yields also a large discrepancy with respect to Ref. [10] for the mean deuterium radius squared. The obtained value of 2.12562(78) fm is six σ smaller than the CODATA value of 2.1413(25) fm. This indicates that the "puzzle" is not limited to the proton.

Table 2.1 Comparison of different experimental values for the RMS radius of the proton

Determination type	$\sqrt{<r_E^2>}$/fm	References
Mainz form factor measurement:	0.879(8)	[8]
CODATA: H and D spectroscopic value: (no *ep* scattering data, no muonic hydrogen)	0.8759(77)	[10]
CODATA: recommended value: (spectroscopy, *e p/e d* scattering data, no muonic hydrogen)	0.8751(61)	[10]
Muonic hydrogen (Lamb shift)	0.84184(67)	[9]

As one combines the measured mean deuterium radius squared of Ref. [11] with the electronic isotope shift to determine a mean proton radius squared of $0.8356(20)$ fm, the value seems to be in agreement with the one obtained from the muonic hydrogen measurement. This even amplifies the "Proton Radius Puzzle", which is still unsolved.

2.2 Deep Inelastic Scattering (DIS)

Deep inelastic lepton nucleon scattering (DIS) is one of the most fundamental tools of high energy physics. Studying the inclusive, semi-inclusive or exclusive cross sections of a lepton l with four-momentum k being scattered of a nucleon N with four-momentum p, allows probing the structure of the nucleon and the interaction mechanisms within. As in the case of the elastic scattering the mediator of the interaction between the lepton and the nucleon is a virtual photon γ^* with four-momentum $q = k - k'$. Effects of the weak interaction are neglected in the following, since the center of mass energy at COMPASS-II of $\sqrt{s} \approx 17.4$ GeV is not sufficient to produce a Z^0 boson.

In contrast to elastic scattering, the final state of an inelastic scattering process consists of more than the scattered lepton l' with four-momentum k' and the recoiled target nucleon with four-momentum p'. It is characterised by the fact that the invariant mass W^2 of the γ^*p system is greater than the mass of the proton M:

$$W^2 c^2 = (q + p)^2 = p^2 + 2pq + q^2 = M^2 c^2 + 2M\nu - Q^2 > M^2 c^2. \qquad (2.3)$$

Looking at Eq. 2.3 several things should be noted, yielding the following definitions of Lorentz invariant inclusive scattering variables:

- The quantity ν is given by:

$$\nu := \frac{pq}{M} \overset{lab}{=} E - E'.$$

Another frequently used variable in this context is:

$$y = \frac{pq}{pk} \overset{lab}{=} \frac{M(E - E')}{ME} = \frac{(E - E')}{E}.$$

- The quantity Q^2 is defined as the negative square of the four-momentum of the virtual photon:

$$Q^2 = -q^2 \overset{lab}{\approx} \frac{4EE'}{c^2} \sin^2\left(\frac{\theta}{2}\right).$$

- Transforming the inequality on the right side of Eq. 2.3 yields:

$$x_{\text{Bj}} := \frac{Q^2}{2M\nu} < 1.$$

The inelasticity of the process is thus characterised by the dimensionales Bjorken scaling variable x_{Bj} being smaller than one. In case of $W^2 = M^2$ it follows from Eq. 2.3 that $x_{\text{Bj}} = 1$, which accounts for the elastic case.

The equations marked with "lab" can be derived by taking into account the definition of the corresponding four-vectors in the laboratory system: $k = (E/c, \vec{k})$, $k' = (E'/c, \vec{k}')$ and $p = (Mc, \vec{0})$, while E and E' denote the initial and respectively final lepton energy and θ the angle between $\vec{k}$ and $\vec{k}'$, the momenta of the in- and outgoing lepton.

With these definitions in mind deep inelastic lepton nucleon scattering can be cha-racterised as the process of lepton nucleon scattering in the limit:

$$Q^2, \nu \to \infty, \quad x_{\text{Bj}} = \text{fixed} < 1.$$

In case of inclusive DIS only the final state lepton is of interest. It should be distinguished from semi-inclusive and exclusive DIS, for which at least one final state hadron or respectively the complete final state is considered. It is often said that in deep inelastic scattering the target nucleon is destroyed and that it fragments into a shower of hadrons. This may be true for most of the processes being considered. But, as it will be discussed in Sect. 2.3, the final state proton may well stay intact, while the creation of an additional real photon or meson accounts for the inequality of Eq. 2.3.

2.2.1 Inclusive DIS

Within the one-photon-exchange approximation the cross section for inclusive deep inelastic scattering can be written as [13, 14]:

$$\frac{d\sigma}{dx_{\text{Bj}}dy} \propto L_{\mu\nu}W^{\mu\nu}$$
$$= \left[L^{(S)}_{\mu\nu}(k, k')W^{\mu\nu(S)}(q, p) + L^{(A)}_{\mu\nu}(k, s_l, k')W^{\mu\nu(A)}(q, p, s_N) \right]. \tag{2.4}$$

The calculation of the cross section is separated into the leptonic tensor L, which describes the electromagnetic interaction at the upper vertex of Fig. 2.2 and the hadronic tensor W, which accounts for the non perturbative QCD structure of the nucleon at the lower "blob" of Fig. 2.2. Within the second line of Eq. 2.4 both tensors have been decomposed into a symmetric (S) and an antisymmetric (A) part. The spin four-vector of the initial lepton l and nucleon N, denoted by s_l and s_N, appear only within the antisymmetric part, which describes the polarised cross section.

Fig. 2.2 Feynman diagram
of the deep inelastic
scattering process [12]

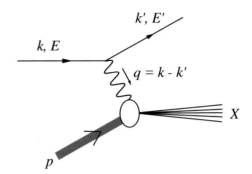

The structure of the hadronic tensor is restricted by symmetry and conservation laws of the strong interaction. Its antisymmetric part can be parametrised by two real functions $g_1(x_{Bj}, Q^2)$ and $g_2(x_{Bj}, Q^2)$, while its symmetric part is given by two real structure functions $F_1(x_{Bj}, Q^2)$ and $F_2(x_{Bj}, Q^2)$.

2.2.2 Unpolarised Inclusive DIS

The unpolarised cross section of lepton nucleon scattering can be parametrised as follows [13, 14]:

$$\frac{d^2\sigma}{dx_{Bj}dy} = \frac{4\pi\alpha^2}{Q^2 x_{Bj}y}\left[x_{Bj}y^2 F_1(x_{Bj}, Q^2) + \left(1 - y - \frac{\gamma^2 y^2}{4}\right)F_2(x_{Bj}, Q^2)\right], \quad (2.5)$$

while γ is given by $\gamma = \frac{2Mx_{Bj}}{Q}$.

Figure 2.3 shows the world data on the experimentally extracted structure function F_2 in dependence of Q^2 and x_{Bj}. In contrast to the elastic scattering cross section, which showed a strong Q^2 dependence, F_2 depends very weakly on Q^2. This was a first hint that pointlike particles are involved in the scattering process, as naively speaking the Fourier transform of a constant function is a δ-distribution. It is this astonishing result, which gave rise to the quark parton model.

In the quark parton model the proton is assumed to be build of pointlike partons, the quarks, antiquarks and gluons. The cross section of Eq. 2.5 can be interpreted as a sum of incoherent elastic lepton quark scattering processes for all possible types of quarks and antiquarks with fractional electric charge e_f. The structure functions F_1 and F_2 can then be expressed in the naive parton model as [15]:

$$F_1(x_{Bj}) = \frac{1}{2}\sum_f e_f^2\left(q_f(x_{Bj}) + \bar{q}_f(x_{Bj})\right), \quad (2.6)$$

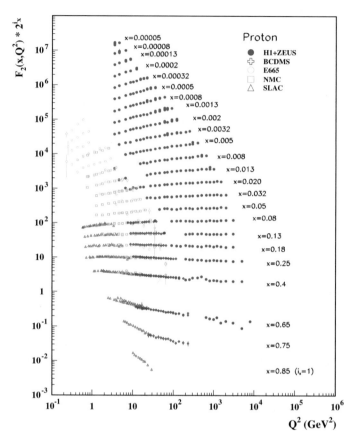

Fig. 2.3 The proton structure function F_2 in dependence of Q^2 and x_{Bj} as extracted by various experiments. For the purpose of plotting F_2^p has been multiplied by $2^{i_x}x$, where i_x denotes the number of the x-bin, ranging from $i_x = 1$ ($x = 0.85$) to $i_x = 24$ ($x = 0.00005$). The corresponding references are given in Ref. [13]

$$F_2(x_{\mathrm{Bj}}) = x_{\mathrm{Bj}} \sum_f e_f^2 \big(q_f(x_{\mathrm{Bj}}) + \bar{q}_f(x_{\mathrm{Bj}})\big). \qquad (2.7)$$

In a fast moving frame with respect to the virtual photon axis the Bjorken variable x_{Bj} can be interpreted as the longitudinal momentum fraction of the parton with respect to the momentum of the proton. The term $(q_f(x_{\mathrm{Bj}})\mathrm{d}x_{\mathrm{Bj}})$ yields the probability of probing a quark of flavour f within the interval $[x_{\mathrm{Bj}}, x_{\mathrm{Bj}} + \mathrm{d}x_{\mathrm{Bj}}]$. This holds likewise for the antiquarks denoted by the "bar" sign.

From Eqs. 2.6 and 2.7 the master equation of the quark parton model is obtained:

$$2x_{\mathrm{Bj}}F_1(x_{\mathrm{Bj}}) = F_2(x_{\mathrm{Bj}}). \qquad (2.8)$$

As in the derivation of Eqs. 2.6 and (2.7) the quarks and antiquarks are assumed to have spin $\hbar/2$, the experimental confirmation of the Callan Gross equation (2.8) [16] confirms the spin-$\frac{1}{2}$ nature of the quarks.

Equations (2.6), (2.7) and (2.8) are valid up to logarithmic corrections in Q^2. The intuitive picture of this so called scaling violation is as follows: With increasing resolving power Q^2 the fact that a quark emits a gluon, which can in turn split into a $q\bar{q}$ pair, is observed. Thus, as Q^2 increases the probability to probe a quark or antiquark with a smaller value of x_{Bj} increases, as it is visible in Fig. 2.3. The precise evolution of the structure functions from one scale, given by Q^2 in this case, to another is given by the DGLAP[2] equations [17–20], while in the kinematic region of very small x_{Bj} it may be more appropriate to sum leading terms in $\ln(1/x_{Bj})$, which is achieved by the so called BFKL[3] equations [21–24].

2.2.3 Longitudinally Polarised Inclusive DIS

In order to determine the polarised structure functions, cross section differences with different target polarisation states are used. In case the incoming lepton is polarised antiparallel to the beam direction $(-)$ and the target is longitudinally polarised either parallel $(+)$ or antiparallel $(-)$ to the beam direction, the cross section difference reads [14]:

$$\frac{d^3\sigma^{-+}}{dx_{Bj}dyd\phi} - \frac{d^3\sigma^{--}}{dx_{Bj}dyd\phi} =$$
$$\frac{4\alpha^2}{Q^2}\left[\left(2 - y - \frac{\gamma^2 y^2}{2}\right)g_1(x_{Bj}, Q^2) - \gamma^2 y^2 g_2(x_{Bj}, Q^2)\right]. \tag{2.9}$$

The structure function g_2 is suppressed by $\frac{1}{Q^2}$, which allows for an almost direct experimental extraction of g_1 with a longitudinally polarised target. In most cases the experimental observable is not the cross section itself, but rather an asymmetry. In this context the directly observable asymmetry $A_{||}$ is given by the cross section difference, according to Eq. 2.9, divided by the unpolarised cross section, according to Eq. 2.5:

$$A_{||} = \frac{d\sigma^{-+} - d\sigma^{-+}}{d\sigma^{-+} + d\sigma^{-+}},$$

while $d\sigma$ is short for $\frac{d^3\sigma}{dx_{Bj}dyd\phi}$. One usually relates $A_{||}$ to the virtual Compton scattering asymmetry A_1 via the optical theorem [15]:

---

[2]**Dokshitzer Gribow Lipatow Altarelli Parisi**.

[3]**Balitskii Fadin Kuraev Lipatov**.

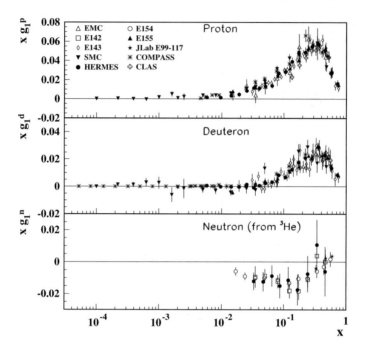

Fig. 2.4 Spin-dependent structure function g_1 of the proton (p), deuteron (d) and neutron (n), extracted with polarised deep inelastic scattering at different fixed target experiments. The corresponding references are given in Ref. [13]

$$A_{\parallel} \approx DA_1 := D\left(\frac{d\sigma_{1/2} - d\sigma_{3/2}}{d\sigma_{1/2} + d\sigma_{3/2}}\right).$$

The quantities $d\sigma_{1/2}$ and $d\sigma_{3/2}$ are the virtual photoabsorption cross sections, in case the projection of the total angular momentum of the $\gamma^* p$ system along the incident lepton direction is $1/2$ or respectively $3/2$. As a virtual photon can have three helicity states the depolarisation factor D, given e.g. in Ref. [15], describes the loss of the incident lepton polarisation due to longitudinal virtual photon polarisation.[4] The asymmetry A_1 has a simple expression in terms of g_1 and g_2 [15]:

$$A_1 = (g_1 - \gamma^2 g_2)/F_1 \approx g_1/F_1. \tag{2.10}$$

Figures 2.4 and 2.5 show the current status of the extraction of g_1.

---

[4]In case the spin and momentum vector of a virtual photon are perpendicular, it is called longitudinally polarised. For historic reasons this is contrary to the usual convention used for massive particles.

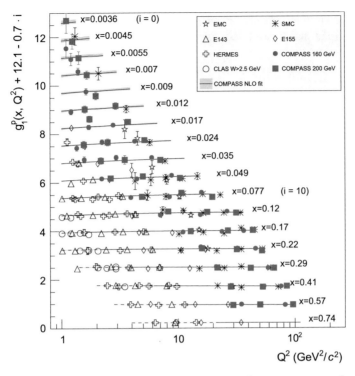

Fig. 2.5 World data on the spin-dependent structure function g_1^p as a function of Q^2 for various values of x_{Bj}. The lines represent the Q^2 dependence for each value of x_{Bj}, as determined from a NLO QCD fit [30]

Within the naive quark parton model $g_1(x_{Bj})$ is given by [15]:

$$g_1(x_{Bj}) = \frac{1}{2} \sum_f e_f^2 \big(\Delta q_f(x_{Bj}) + \Delta \bar{q}_f(x_{Bj}) \big), \qquad (2.11)$$

while the helicity distribution of a quark with flavour f is denoted by:

$$\Delta q_f(x_{Bj}) = q_f^{\rightarrow}(x_{Bj}) - q_f^{\leftarrow}(x_{Bj}).$$

The polarised parton distribution functions $q_f^{\rightarrow}(x_{Bj})$ and $q_f^{\leftarrow}(x_{Bj})$ denote the probability densities to probe a quark with same and respectively opposite spin direction with respect to the longitudinally polarised nucleon. They are also logarithmically dependent on Q^2 with the same remarks being valid as for the unpolarised case.

A particular intriguing quantity in spin physics is the first moment of g_1. It is given in leading order by [25]:

$$\int_0^1 g_1(x_{Bj}, Q^2) = \frac{1}{12}\left(a_3 + \frac{1}{3}a_8\right) + \frac{1}{9}a_0, \qquad (2.12)$$

and is linked to the isovector charge a_3, the octed charge a_8 and the flavour-singlet charge a_0. In terms of flavour composition a_3 and a_8 are given for the proton by:

$$a_3 = \Delta u - \Delta d + (a.q.), \qquad a_8 = \Delta u + \Delta d - 2\Delta s + (a.q.), \qquad (2.13)$$

while the abbreviation $(a.q)$ denotes the same terms for the corresponding antiquarks and the notation:

$$\Delta q_f = \int_0^1 \Delta q_f(x_{Bj})dx_{Bj},$$

is used. The isovector charge a_3 is equal to the weak coupling constant $|\frac{g_A}{g_V}|$, while the octed charge a_8 is known from hyperon decay and the assumption of SU(3) flavour symmetry. The contribution of the quarks and antiquarks to the spin of the nucleon is given by:

$$a_0 = \Delta\Sigma = \sum_f \Delta q_f + (a.q.). \qquad (2.14)$$

For higher orders in QCD $\Delta\Sigma$ becomes Q^2 dependent and Eq. 2.12 is modified by corrections at the order of the strong coupling constant. In the $\overline{MS}$ renormalisation scheme the singlet axial charge $a_0(Q^2)$ is still identical to $\Delta\Sigma(Q^2)$ [26], while in the off shell scheme it is shown that the gluon polarisation Δg can also contribute to a_0 [27]:

$$a_0(Q^2) = \Delta\Sigma_{off} - 3\frac{\alpha_s(Q^2)}{2\pi}\Delta g(Q^2). \qquad (2.15)$$

A historic measurement was the EMC result for the first moment of g_1 for the proton. Using Eq. 2.12 and (2.13) together with the measured moment over g_1 and the known values of a_3 and a_8, the EMC deduced a singlet axial charge a_0 compatible with zero and a sizeable negative strange quark contribution $(\Delta s + \Delta\bar{s})$ [28]. The identification of a_0 with $\Delta\Sigma$, using Eq. 2.14, lead to the conclusion that a negligible amount of the proton spin originates from the quarks and antiquarks.

These findings are in strong contrast to the static quark parton model, which predicts that the proton spin originates solely from the spins of the valence quarks. Even in relativistic parton models a contribution of 60% of the quarks and antiquarks is expected.

While the value of a_0 is somewhat larger with today's knowledge and at the order of 0.3, the basic conclusions of the EMC stay unchanged. In particular, the sizeable negative contribution of the strange quarks, which is historically related to the breaking of the Ellis–Jaffe sum rule [29], could be confirmed within modern inclusive DIS experiments (see e.g. Refs. [30, 31]).

2.2.4 Longitudinally Polarised Semi-Inclusive DIS

To shed more light on the decomposition of the proton spin, semi-inclusive measurements have to be performed. These measurements allow for a somehow direct access to the individual helicity distributions of quarks, antiquarks and gluons, In fact, in some sense Eq. 2.15 gave birth to the COMPASS experiment. COMPASS was expected to measure a large contribution of Δg at the order of 2-3, which was believed to mask the true value of $\Delta\Sigma$.

Flavour Specific Helicity Distributions

It is worth recalling the expression of the inclusive asymmetry A_1 by using Eq. 2.10 together with the expressions (2.11) and (2.6) of g_1 and F_1 in the simple parton model:

$$A_1(x_{\text{Bj}}) = \frac{\sum_f e_f^2 \big(\Delta q_f(x_{\text{Bj}}) + \Delta\bar{q}_f(x_{\text{Bj}})\big)}{\sum_f e_f^2 \big(q_f(x_{\text{Bj}}) + \bar{q}_f(x_{\text{Bj}})\big)}, \tag{2.16}$$

In complete analogy to Eq. 2.16 one defines the asymmetry [32]:

$$A_1^h(x_{\text{Bj}}, z) = \frac{\sum_f e_f^2 \big(\Delta q_f(x_{\text{Bj}})D_f^h(z) + \Delta\bar{q}_f(x_{\text{Bj}})D_{\bar{f}}^h(z)\big)}{\sum_f e_f^2 \big(q_f(x_{\text{Bj}})D_f^h(z) + \bar{q}_f(x_{\text{Bj}})D_{\bar{f}}^h(z)\big)}, \tag{2.17}$$

in case a hadron h is observed in addition to the scattered muon. It is valid in the leading order QCD parton model under the assumption of independent quark fragmentation and in case the hadrons are produced in the current fragmentation region [32]. The fragmentation functions $D_f^h(z)$ and $D_{\bar{f}}^h(z)$ are quite similar to the parton distribution functions. But they describe in turn the probability that the struck quark or antiquark of flavour f fragments into a hadron h with energy E_h, carrying the energy fraction $z = E_h/\nu$ of the struck quark.

Using the fragmentation functions and parton distribution functions extracted for example from unpolarised semi-inclusive deep inelastic scattering experiments, one can disentangle the parton specific helicity distributions from Eq. 2.17. The reason one gains sensitivity to the individual helicity distributions is easy to understand since for example an observation of a kaon in the final state directly points to the fact that the struck quark was most likely an s quark. A similar simple intuition can be gained for charged pions, as a π^+ is more likely originating from an up quark than a π^- and vice versa for the down quark.

Figure 2.6 shows a recent leading order extraction of the quark helicity distributions at the COMPASS experiment. It is interesting to note that the helicity distribution of the strange quarks is compatible with zero. This is in contrast to the x_{Bj} integrated inclusive determination described in the last section and explains the recent efforts in the validation of the kaon fragmentation functions [33]. In Ref. [32] two values for $\Delta\Sigma$ at $Q_0^2 = 3\,(\text{GeV/c})^2$ are given:

$$\Delta\Sigma_{extrap} = 0.32 \pm 0.03 \pm 0.03, \quad \Delta\Sigma_{DSSV} = 0.22 \pm 0.03 \pm 0.03.$$

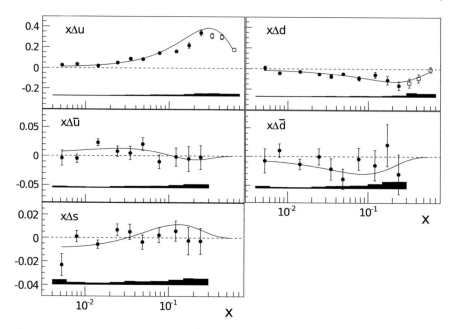

Fig. 2.6 Quark helicity distributions at $Q^2 = 3(\text{GeV/c})^2$ [32]

The value for $\Delta\Sigma_{extrap}$ is extracted by using a linear interpolation of the data for the x_{Bj} integration, while $\Delta\Sigma_{DSSV}$ uses the DSSV [34, 35] parametrisation of the helicity distributions. The disagreement is again tracked down to the strange quark helicity and still under investigation. In any case, regardless of the strange quark helicity, this independent determination of $\Delta\Sigma$ supports the qualitative statement of the EMC, regarding the small quark and antiquark contribution to the spin of the proton.

The Gluon Helicity Distribution

The gluon helicity can not be probed in DIS via the leading order virtual photon absorption, shown at the left side of Fig. 2.7, because there is no direct coupling of the virtual photon to the gluon. The higher order process of photon-gluon-fusion, shown at the right side of Fig. 2.7, gives access to the gluon helicity though. A very clean way to gain sensitivity to the photon-gluon-fusion process is via so called open charm production. The selection of two charmed mesons such as D^0 and $\overline{D^0}$ in the final state is an almost direct experimental signature of the photon-gluon-fusion since for most kinematics the charm content in the nucleon is negligible and the production of charmed mesons within the fragmentation from light quarks is highly suppressed. The downside is that due to the large mass of the charm quark the production of charmed mesons within the photon-gluon-fusion process is highly suppressed, which leads to the very limited statistical accuracy in this channel. For more details on the open charm production at COMPASS it shall be referred to Ref. [36].

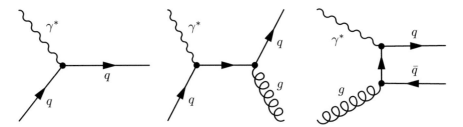

Fig. 2.7 Feynman diagrams for virtual photon nucleon scattering. Left: Leading order process (LP). Middle: Gluon radiation respectively QCD Compton scattering (QCDC). Right: Photon-gluon-fusion (PGF). Picture adopted from Ref. [37]

The question of how to increase the statistical accuracy in probing the gluonic content of the nucleon lead to the idea that the requirement of high transverse momenta of two final state hadrons also enhances the sensitivity on the photon-gluon-fusion process. The transverse momentum of hadron pairs produced in the leading order virtual photon absorption is mainly originating from the intrinsic transverse momenta of the quarks in the nucleon together with the transverse momenta produced within the fragmentation process. In case of the photon-gluon-fusion process and the QCD Compton scattering process, shown in the middle of Fig. 2.7, the transverse momenta of the final state hadrons mainly originate from the hard process and are supposed to be significantly larger. The sensitivity of the extraction of the gluon helicity thus relies on the distinct behaviour of the transverse momenta of the final state hadrons between the leading order virtual photon absorption and the two higher order processes. At leading order in QCD and under the assumption of spin independent fragmentation the experimentally observable longitudinal double spin asymmetry is given by [37]:

$$A_{||}^{2h}(x_{\mathrm{Bj}}) = R_{PGF} a_{||}^{PGF} \frac{\Delta g}{g}(x_g) + R_{LP} D A_1^{LP}(x_{\mathrm{Bj}}) + R_{QCDC} a_{||}^{QCDC} A_1^{LP}(x_C).$$

The quantity A_1^{LP} is given at leading order by Eq. 2.16. The quantities $R_{\{PGF,LP,QCDC\}}$ are the fractions of the corresponding process illustrated in Fig. 2.7 and are usually estimated by Monte Carlo techniques. The quantities $a_{||}^{PGF}$ and $a_{||}^{QCDC}$ are the asymmetries of the partonic cross section, which are often referred to as analysing power [38]. In case of the inclusive asymmetry A_1^{LP} the analysing power is given by the depolarisation factor D introduced in Sect. 2.2.1. The variables x_{Bj}, x_g and x_C denote the quark momentum fraction, the gluon momentum fraction and the quark momentum fraction in the QCD Compton scattering process. Though further peculiarities have to be considered, which are related to the fact that the QCD Compton scattering process and the photon-gluon-fusion process are also present within A_1^{LP} and that the variables x_g and x_c are not directly accesible, the principle knowledge of the inclusive asymmetry A_1^{LP} together with the fractions of the corresponding processes and the analysing powers allow for an extraction of $\frac{\Delta g}{g}(x_g)$.

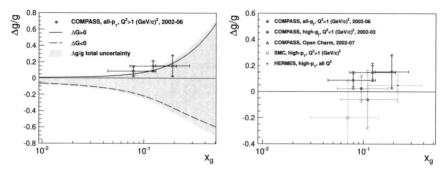

Fig. 2.8 Left: Comparison of the leading order results for $\Delta g/g(x_g)$ with a COMPASS NLO QCD fit [30]. Right: World data on $\Delta G/G(x_g)$. The corresponding references are given within Ref. [39]

Figure 2.8 shows a recent extraction of $\frac{\Delta g}{g}(x_g)$, which seems to favour a positive but small value for Δg in case solely the COMPASS data is used. Recent results from RHIC[5] confirm the small contribution of Δg, but they may indicate that there is still a sizeable contribution of the gluon to the spin of the nucleon. This is mainly related to the poorly known region of $x_{Bj} < 0.05$ [40]. In any case it seems very unlikely with the current knowledge that the contribution of Δg within Eq. 2.15 is sizeable enough to explain the "small" value for $\Delta \Sigma$.

2.3   Generalised Parton Distributions

One of the many theoretical attempts to explain the "small" experimental value of $\Delta \Sigma$ was the spin decomposition proposed by Jaffe and Manohar [41]:

$$\frac{1}{2} = \frac{1}{2}\Delta \Sigma + \Delta g + L_q + L_g.$$

From this it became evident that a comprehensive picture of the spin of the nucleon must take into account the orbital angular momentum L_q of quarks and antiquarks together with the orbital angular momentum L_g of the gluons. As there are measurements of $\Delta \Sigma$ and Δg, discussed in Sect. 2.2, there is no experimental prescription so far of how to access the contribution originating from the orbital angular momenta of the partons.

In 1997 a completely independent and comprehensive approach to the spin decomposition of the nucleon was proposed. The Ji sum rule [42]:

---

[5]**R**elativistic **H**eavy **I**on **C**ollider: RHIC performs polarised *pp* collisions in Brookhaven.

$$J^f = \frac{1}{2} \lim_{t \to 0} \int_{-1}^{1} \left[H^f(x, \xi, t) + E^f(x, \xi, t) \right] x \, dx,$$

$$J^g = \frac{1}{2} \lim_{t \to 0} \int_{0}^{1} \left[H^g(x, \xi, t) + E^g(x, \xi, t) \right] dx,$$

(2.18)

connects so called Generalised Parton Distributions H and E to the total angular momentum of gluons g and quarks of flavour f. It is this relation, which triggered a lot of the experimental and theoretical interest in Generalised Parton Distributions during the following years. The following chapter will define the kinematic variables used in Eq. 2.18 and summarise the current knowledge on Generalised Parton Distributions.

2.3.1 Introduction

Generalised Parton Distributions (GPDs) provide a comprehensive three dimensional picture of the nucleon, encoded in their dependence on the three kinematic variables x, ξ and t and a weak dependence on Q^2 describing the QCD evolution. It is most illustrative to explain the kinematic variables in the picture of a particular process. Figure 2.9 shows a so called handbag diagram for the Deeply Virtual Compton scattering process (DVCS). In the Bjorken limit the process can be factorised into a hard and a soft part in case the ratio of the magnitude of the square of the four-momentum transfer to the proton and the photon virtuality, $|t|/Q^2$, is sufficiently small [44].

The hard part consists of a quark carrying longitudinal momentum fraction $x + \xi$, which interacts with the virtual photon and returns into the nucleon with longitudinal momentum fraction $x - \xi$ under the emission of a real photon at a different transverse position. In this context the variable x is a loop variable, describing the momentum fraction carried by the quark with respect to the mean longitudinal momentum of the nucleon throughout the process. It is not accessible within the measurement. The variable ξ is related to x_{Bj} by [45, 46]:

Fig. 2.9 Handbag diagram for the DVCS process [43]

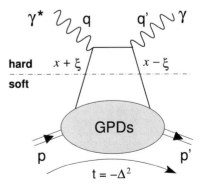

$$\xi = x_{\text{Bj}} \frac{1 + \frac{\Delta^2}{2Q^2}}{2 - x_{\text{Bj}} + x_{\text{Bj}} \frac{\Delta^2}{Q^2}} \approx \frac{x_{\text{Bj}}}{2 - x_{\text{Bj}}}.$$

The finite four-momentum transfer necessary to force the virtual photon on its mass shell is given by:

$$t = (p - p')^2 = -\Delta^2,$$

while p and p' denote the four momenta of the initial and final nucleon.

The soft part is given by the emission and reabsorption of the quark within the nucleon. It can be parametrised by four process independent non perturbative objects for the gluon g and each quark flavour f, the GPDs $H^{f,g}$, $\tilde{H}^{f,g}$, $E^{f,g}$, $\tilde{E}^{f,g}$. The GPDs H and E do not depend on the helicity of the struck quark, while the GPDs $\tilde{H}$ and $\tilde{E}$ are dependent on the quark helicity. The latter two can thus be probed most effectively with a polarised target. The nucleon helicity is conserved by the GPDs H and $\tilde{H}$, while it is flipped by the GPDs E and $\tilde{E}$.

As the quantities x and ξ are limited to the interval $x \in [-1, 1]$ and $\xi \in [0, 1]$, there are in principle two cases to distinguish:

For $x \in [-\xi, \xi]$ the momentum transfer $x + \xi$ is positive, while the momentum transfer $x - \xi$ is negative. This so called ERBL region[6] corresponds to the emission of a quark antiquark pair. There is no correspondence in the forward limit, when $\xi = 0$, and in this case GPDs behave rather like a meson distribution amplitude and can be interpreted as the probability amplitude to find a quark antiquark pair within the nucleon.

In case x lies in the interval $[\xi, 1]$ ($[-1, -\xi]$) both momentum fractions $x + \xi$ and $x - \xi$ are positive (negative) and the GPDs describe the emission and reabsorption of a quark (antiquark), as it is shown in Fig. 2.9. This is commonly referred to as the DGLAP region and there is a correspondence to the usual parton distribution functions in the forward limit

2.3.2 Forward Limit

In the forward limit, defined by the condition:

$$t \to 0 \text{ and } \xi \to 0,$$

the GPDs H and $\tilde{H}$ are related to the ordinary parton distribution functions as follows [46]:

---

[6]**E**fremov, **R**adyushkin, **B**rodsky, **L**epage region: The term originates from the corresponding ERBL evolution equations [47, 48].

$$H^f(x,0,0) = q(x), \qquad \tilde{H}^f(x,0,0) = \Delta q(x) \qquad \text{for } x > 0,$$
$$H^f(x,0,0) = -\bar{q}(-x), \qquad \tilde{H}^f(x,0,0) = \Delta\bar{q}(-x) \qquad \text{for } x < 0,$$
$$H^g(x,0,0) = xg(x), \qquad \tilde{H}^g(x,0,0) = x\Delta g(x) \qquad \text{for } x > 0, \qquad (2.19)$$

while in this limit x coincides with x_{Bj}. For the GPDs E and $\tilde{E}$ there is no relation to the parton distribution functions in the forward limit as they describe a nucleon helicity flip, which is not possible for a vanishing four-momentum transfer of the nucleon. They contain unique information about the spin of the nucleon (see Eq. 2.18), which is only accessible within exclusive processes.

2.3.3 Sum Rules

The most popular sum rule for GPDs has already been stated in the introduction in Eq. 2.18. Furthermore, the first moments of GPDs are linked to the elastic Form Factors [42]:

$$\sum_f z_f \int_{-1}^{1} dx H^f(x,\xi,t) = F_1(t), \qquad \sum_f z_f \int_{-1}^{1} dx \tilde{H}^f(x,\xi,t) = g_A(t),$$

$$\sum_f z_f \int_{-1}^{1} dx E^f(x,\xi,t) = F_2(t), \qquad \sum_f z_f \int_{-1}^{1} dx \tilde{E}^f(x,\xi,t) = h_A(t). \qquad (2.20)$$

The quantities g_A and h_A denote the axial and pseudoscalar Form Factors, while the Dirac and Pauli Form Factors F_1 and F_2 are discussed in Sect. 2.1. The GPDs thus describe the contribution to the corresponding Form Factor for a given mean longitudinal momentum fraction x. It is quite revealing that the ξ dependence drops out in the Eq. 2.20, as the integration over x removes all reference to the longitudinal direction, which is used for the definition of ξ. In Ref. [42] X. Ji even used the term "luckily" related to this fact.

One can consider even higher moments in x. This leads to the so called polynomiality feature of GPDs, which states that the n-th moment of the GPDs are polynomials in ξ maximally of the order $n + 1$. For the quark GPDs H^f and E^f it reads [46]:

$$\int_{-1}^{1} dx\, x^n H^f(x,\xi,t) = \begin{cases} a_0^n(t) + a_2^n(t)\xi^2 + a_4^n(t)\xi^4 + \ldots + a_n^n(t)\xi^n, & n \text{ even}, \\ a_0^n(t) + a_2^n(t)\xi^2 + a_4^n(t)\xi^4 + \ldots + c_{n+1}^f(t)\xi^{(n+1)}, & n \text{ odd}, \end{cases}$$

and

$$\int_{-1}^{1} dx\, x^n E^f(x,\xi,t) = \begin{cases} b_0^n(t) + b_2^n(t)\xi^2 + b_4^n(t)\xi^4 + \ldots + b_n^n(t)\xi^n, & n \text{ even}, \\ b_0^n(t) + b_2^n(t)\xi^2 + b_4^n(t)\xi^4 + \ldots - c_{n+1}^f(t)\xi^{(n+1)}, & n \text{ odd}. \end{cases}$$

It originates from the time reversal invariance that only even powers in ξ appear [46]. The relations look similar for the GPDs $\tilde{H}^f$ and $\tilde{E}^f$, apart from the fact that the highest power in ξ is given by n in case it is an even number and by $(n-1)$ in case it is an odd number. Furthermore, neither of the coefficients would cancel if one takes the sum of the n-th moments of $\tilde{H}^f$ and $\tilde{E}^f$.

In case of the gluon GPDs H^g and E^g the above relations for the $(n-1)$-th moment read [46]:

$$\int_0^1 dx\, x^{n-1} H^f(x, \xi, t) = \begin{cases} 0, & n \text{ even,} \\ d_0^n(t) + d_2^n(t)\xi^2 + d_4^n(t)\xi^4 + \ldots + c_{n+1}^g(t)\xi^{(n+1)}, & n \text{ odd,} \end{cases}$$

and

$$\int_0^1 dx\, x^{n-1} E^f(x, \xi, t) = \begin{cases} 0, & n \text{ even,} \\ e_0^n(t) + e_2^n(t)\xi^2 + e_4^n(t)\xi^4 + \ldots - c_{n+1}^g(t)\xi^{(n+1)}, & n \text{ odd,} \end{cases}$$

with the same remarks being valid as for the quark GPDs, with the exception that in case of n being an even number, the moments of $\tilde{H}^g$ and $\tilde{E}^g$ vanish. The fact that the $(n-1)$-th moments of H^g and E^g vanish for n being odd and the ones of $\tilde{H}^g$ and $\tilde{E}^g$ for n being even, is related to the symmetry properties of the gluon GPDs. Since the gluon is its own antiparticle, H^g and E^g are even functions in x while $\tilde{H}^g$ and $\tilde{E}^g$ are odd in x.

The relations (2.19) and (2.20) provide valuable constraints for GPD models, while the polynomiality feature allows to restrict the class of usable functions within a particular model. A very elegant way to satisfy polynomiality is the so called double distribution ansatz [49–51]. It was observed though that within the double distribution ansatz the coefficients c_{n+1}^f and respectively c_{n+1}^g always vanish. This incompleteness of the double distribution ansatz then lead to the introduction of the so called D-term [52], which is added to the double distribution ansatz to generate the highest power of ξ for the moments of H, E and n being odd.

2.3.4 Impact Parameter Space

In Sect. 2.3.1 GPDs have been introduced in momentum space. A very intuitive three dimensional picture of the nucleon arises in the so called mixed representation of longitudinal momentum and transverse position. In case $\xi = 0$, the longitudinal momentum fraction of the quark in the initial and final state is equal and the four-momentum transfer to the nucleon is aligned purely in the transverse direction $\Delta^2 = \Delta_\perp^2$. In this particular situation it is shown that the Fourier transform of the GPD H with respect to $\Delta_\perp$ has a density interpretation. The quantity:

$$q_f(x, \mathbf{b}_\perp) = \int \frac{d^2\Delta_\perp}{(2\pi)^2} H_q(x, 0, -\Delta_\perp^2) e^{-i\mathbf{b}_\perp \Delta_\perp},$$

gives the probability density to probe a quark with longitudinal momentum fraction x at the transverse distance $b_\perp$ with respect to

$$\mathbf{R}_\perp = \sum_i x_i \, \mathbf{r}_{\perp,i},$$

the centre of momentum of the nucleon in the transverse plane [53]. The longitudinal momentum fractions of the partons are denoted by x_i, while i runs over all partons in the nucleon. Here and in the following two dimensional transverse vectors are written in bold face, while three dimensional vectors are indicated by an arrow. The transverse centre of momentum $\mathbf{R}_\perp$ plays the role of the centre-of-mass in a nonrelativistic many body system, with masses m_i corresponding to the longitudinal momentum fractions x_i.

This interpretation of GPDs plays an import role in the modelling of GPDs at $\xi = 0$. For illustration purposes Fig. 2.10 shows a model ansatz for the GPD H:

$$H_q(x, 0, -\Delta_\perp{}^2) = q(x) \exp\left(-a\Delta_\perp^2 (1 - x) \ln(1/x)\right), \qquad (2.21)$$

transformed to the impact parameter space:

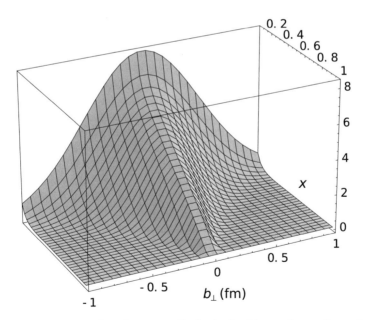

Fig. 2.10 Impact parameter dependent parton distribution for the u quark according to the simple model following Eq. 2.21 and respectively (2.22). Picture adopted from Ref. [54]

$$q(x, \mathbf{b}_\perp) = q(x) \frac{1}{4\pi a(1-x)\ln(1/x)} \exp\left(-\frac{\mathbf{b}_\perp^2}{4a(1-x)\ln(1/x)}\right). \qquad (2.22)$$

This ansatz is in agreement with several facts. The integration over x evaluated at $\Delta_\perp^2 = 0$ of $H_q(x, 0, -\Delta_\perp^2)$ yields the parton distribution function $q(x)$. The transverse width of the impact parameter version $q(x, \mathbf{b}_\perp)$ converges to zero as $x \to 0$, which is in agreement with the fact that for $x = 0$ the transverse centre of momentum $\mathbf{R}_\perp$ is given by the struck quark alone. Last but not least as expected from a density $q(x, \mathbf{b}_\perp)$ satisfies the relation:

$$q(x, \mathbf{b}_\perp) \geq 0 \text{ for all } x > 0.$$

In practice it is not possible to measure GPDs at $\xi = 0$, as it will be demonstrated within the example of the DVCS process in Sect. 2.4. Thus, the density interpretation can not be applied directly to the measured data. Though theoretical constraints, like in particular the polynomiality feature of Sect. 2.3.3, facilitate the extrapolation to $\xi = 0$, it is still almost impossible to quantify the model uncertainties introduced within this extrapolation.

From an abstract point of view it is quite easy to understand why the above interpretation can not be extended to non zero skewness ξ. GPDs are defined as transition matrix elements. In order to provide a density interpretation the initial and final state have to coincide. For the case $\xi = 0$ the longitudinal momenta of the initial and final state already coincide. Hence, the main task in order to provide the density interpretation of Ref. [53] is to show that the Fourier transformation with respect to the transverse momentum transfer also yields identical initial and final states in terms of transverse position. This is achieved by introducing the transverse centre of momentum in close analogy to the centre of mass being a conserved quantity in the nonrelativistic case. However, for a finite longitudinal momentum transfer ξ the fact that the initial and final longitudinal momenta of the nucleon do not coincide can not be overcome. This restricts the density interpretation to the case of $\xi = 0$.

Nevertheless, proceeding in this direction, within Ref. [55] it is shown that in the case $\xi \neq 0$ also in the impact parameter space the initial and final states are not equal. As the struck quark looses part of its longitudinal momentum, the transverse centre of momentum is shifted between the initial and final state by an amount of order $\xi \mathbf{b}_\perp$, as illustrated in Fig. 2.11. Since the four-momentum transfer is not purely transverse in the case of $\xi \neq 0$, the quantity $\mathbf{b}_\perp$ is the Fourier conjugate to the transverse part of Δ given by $\Delta_\perp$ according to [56][7]:

$$\Delta^2 = -t_0 + \frac{1+\xi}{1-\xi}\Delta_\perp^2 = \frac{4\xi^2 M^2}{(1-\xi)(1+\xi)} + \frac{1+\xi}{1-\xi}\Delta_\perp^2. \qquad (2.23)$$

The quantity t_0 refers to the minimum value of the square of the four-momentum transfer to the nucleon. Though it is argued that for small ξ the shift in the trans-

---

[7]See Eq. (13) in [57] and use $\zeta = \frac{2\xi}{1+\xi}$ and $t_0 = -\frac{\zeta^2 M^2}{1-\zeta}$.

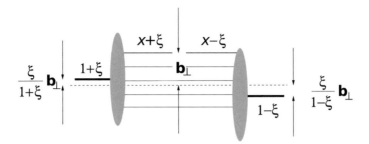

Fig. 2.11 Representation of a GPD in the impact parameter space for the region $\xi \le x \le 1$. Picture adapted from Ref. [55]

verse plane is almost irrelevant, one can change to the centre of momentum of the spectators. This centre of momentum is conserved throughout the process because it is not connected to the struck quark. The distance of the struck quark to the centre of momentum of the spectators $\mathbf{r}_\perp$ is Fourier conjugate to $\mathbf{b}_\perp$ in the particular interesting case of $x = \xi$ [56]. Thus, assuming the transition matrix element behaves exponential as a function of $\Delta_\perp^2$ with a slope $B_{\Delta_\perp^2}$, its transverse size is given by:

$$< r_\perp^2 >= 4B_{\Delta_\perp^2} = 4\left(\frac{1+\xi}{1-\xi}\right)B_t. \tag{2.24}$$

The second equality in Eq. 2.24 refers to the fact that usually in measurements the so called t-slope parameter B_t is reported, which parametrises the $|t|$ dependence of the DVCS cross section. The relation between the two slopes $B_{\Delta_\perp^2}$ and B_t arises from Eq. 2.23.

To summarise, measurements at $x = \xi$ allow studying the transverse size of the transition matrix element, defining GPDs, as a function of the longitudinal momentum fraction of the struck quark. This is often referred to as nucleon tomography.

2.4 Deeply Virtual Compton Scattering

Deeply Virtual Compton Scattering (DVCS):

$$l + N \to l' + N' + \gamma,$$

describes the scattering of a high energy lepton of the nucleon via the exchange of a virtual photon in the limit:

$$Q^2, \nu \to \infty, \ x_{\mathrm{Bj}} = \text{fixed}, \ |t|/Q^2 < 1.$$

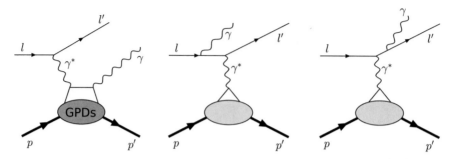

Fig. 2.12 Leading-order processes for leptoproduction of real photons. Left: DVCS. Middle and right: Bethe-Heitler (BH) process. Picture adopted from Ref. [54]

It can be accessed within exclusive measurements. Thus, for a clean experimental signature it is mandatory to also detect the recoiled target nucleon, apart from the scattered lepton and the real photon.

The DVCS process offers a way to experimentally constrain GPDs, which parametrise the soft part of the left diagram in Fig. 2.12, as described in Sect. 2.3.1. It is the most pure channel to study GPDs since in contrast to the hard exclusive production of a meson no final state interaction and no meson wave function have to be taken into account.

However, DVCS is not the only process, which describes the reaction 1.4. The initial and final states of the Bethe-Heitler process, illustrated in the middle and on the right side of Fig. 2.12, are indistinguishable from DVCS. The Bethe–Heitler process describes elastic scattering of the lepton of the nucleon, while both the incoming and outgoing lepton can emit a real photon.

The two processes interfere on the amplitude level and the differential cross section can be written schematically as [45, 54]:

$$\frac{\mathrm{d}^4\sigma}{\mathrm{d}x_{\mathrm{Bj}}\,\mathrm{d}Q^2\,\mathrm{d}|t|\,\mathrm{d}\phi_{\gamma*\gamma}} \propto |\mathcal{T}_{BH}|^2 + |\mathcal{T}_{DVCS}|^2 + \mathcal{I},$$

with:

$$\mathcal{I} = \mathcal{T}_{BH}^*\mathcal{T}_{DVCS} + \mathcal{T}_{BH}\mathcal{T}_{DVCS}^*.$$

The angle $\phi_{\gamma*\gamma}$ denotes the angle between the leptonic plane and the plane spanned by the real and the virtual photon, as illustrated in Fig. 2.13. The complex scattering amplitudes of the respective process are depicted by $\mathcal{T}$.

Changing the charge and polarisation of the lepton beam and using unpolarised, longitudinally or transversely polarised proton or deuteron targets, a variety of experimental observables such as cross section differences, sums and asymmetries of the different configurations can be accessed within DVCS. A complete description of the theoretical formalism, which provides the connection between the different observables within different experimental setups and GPDs is given in Ref. [45], while the two cases which are of particular interest for the COMPASS-II experiment shall be discussed in the following.

Fig. 2.13 Definition of the angle $\phi_{\gamma*\gamma}$. Picture adopted from [58]

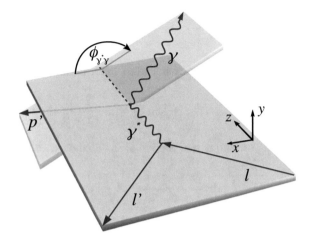

The COMPASS-II experiment has the unique feature to change simultaneously the charge and polarisation of the muon beam. Within the recent DVCS measurements an unpolarised liquid hydrogen target is used. The differential cross section reads in this case [43]:

$$\frac{\mathrm{d}^4\sigma}{\mathrm{d}x_{\mathrm{Bj}}\mathrm{d}Q^2\mathrm{d}|t|\mathrm{d}\phi_{\gamma*\gamma}} = \mathrm{d}\sigma^{\mathrm{BH}} + \left(\mathrm{d}\sigma_{unpol}^{\mathrm{DVCS}} + P_\mu \mathrm{d}\sigma_{pol}^{\mathrm{DVCS}}\right) + e_\mu\left(\mathrm{Re}I + P_\mu\mathrm{Im}I\right).$$

(2.25)

On the right side of Eq. 2.25 and in the following the abbreviation:

$$\mathrm{d}\sigma = \frac{\mathrm{d}^4\sigma}{\mathrm{d}x_{\mathrm{Bj}}\mathrm{d}Q^2\mathrm{d}|t|\mathrm{d}\phi_{\gamma*\gamma}},$$

will be used. The polarisation and charge in units of the elementary charge e are denoted by P_μ and e_μ respectively and the remaining terms are defined in the following sections.

The main observables of interest in case of the COMPASS-II DVCS programme are thus given by the unpolarised beam charge and spin sum or difference of DVCS cross sections, which will be discussed in the following, after a short introduction to the subject of Compton Form Factors has been given.

2.4.1 Compton Form Factors

The variable x, describing the mean longitudinal momentum fraction of the struck quark throughout the process, can not be accessed directly by a measurement of DVCS. This fact is encoded in so called Compton Form Factors. A Compton Form Factor is connected to the respective GPD via a convolution integral in x, taking into account the hard scattering kernel, originating from the virtual photon quark interaction. In case of the Compton Form Factor $\mathcal{H}$ the relation explicitly reads [45]:

$$\mathcal{H}(x, \xi, t) = \sum_f e_f^2 \int_{-1}^{1} dx \, C_c^-(x, \xi) H^f(x, \xi, t),$$

while $C_c^\pm(x, \xi)$ is given in leading order of the strong coupling constant by [45]:

$$C_c^\pm(x, \xi) = \frac{1}{\xi - x - i\epsilon} \pm \frac{1}{\xi + x - i\epsilon}.$$

Making use of the real version of the Sokhotski-Plemelj theorem:

$$\int_{-1}^{1} dx \, \frac{H^f(x, \xi, t)}{\xi \pm x - i\epsilon} = \mathcal{P} \int_{-1}^{1} dx \, \frac{H^f(x, \xi, t)}{\xi \pm x} + i\pi H^f(\mp\xi, \xi, t),$$

the Compton Form Factor $\mathcal{H}$ can be divided into its real and imaginary part:

$$\mathcal{H} = \sum_f e_f^2 \left[\mathcal{P} \int_{-1}^{1} dx \, H^f(x, \xi, t) C^-(x, \xi) + i\pi \Big(H^f(\xi, \xi, t) - H^f(-\xi, \xi, t) \Big) \right],$$

while $\mathcal{P}$ denotes a principal value integral and:

$$C^\pm(x, \xi) = \frac{1}{\xi - x} \pm \frac{1}{\xi + x}, \tag{2.26}$$

has been introduced. In many applications the integration over x is converted to the interval $[0, 1]$, which leads to the connection of the four so called singlet GPDs, denoted by the subscript $+$:

$$\{H_+, E_+\}(x, \xi, t) = \sum_f e_f^2 \Big(\{H^f, E^f\}(x, \xi, t) - \{H^f, E^f\}(-x, \xi, t) \Big),$$

$$\{\tilde{H}_+, \tilde{E}_+\}(x, \xi, t) = \sum_f e_f^2 \Big(\{\tilde{H}^f, \tilde{E}^f\}(x, \xi, t) + \{\tilde{H}^f, \tilde{E}^f\}(-x, \xi, t) \Big), \tag{2.27}$$

to the real and imaginary parts of the four Compton Form Factors:

$$\{\mathcal{H}_{Re}, \mathcal{E}_{Re}\}(\xi, t) = \mathcal{P} \int_0^1 dx \, \{H_+, E_+\}(x, \xi, t) \, C^-(x, \xi),$$

$$\{\tilde{\mathcal{H}}_{Re}, \tilde{\mathcal{E}}_{Re}\}(\xi, t) = \mathcal{P} \int_0^1 dx \, \{\tilde{H}_+, \tilde{E}_+\}(x, \xi, t) \, C^+(x, \xi),$$

$$\{\mathcal{H}_{Im}, \mathcal{E}_{Im}\}(\xi, t) = \pi \{H_+, E_+\}(\xi, \xi, t),$$

$$\{\tilde{\mathcal{H}}_{Im}, \tilde{\mathcal{E}}_{Im}\}(\xi, t) = \pi \{\tilde{H}_+, \tilde{E}_+\}(\xi, \xi, t). \tag{2.28}$$

The imaginary parts of the Compton Form Factors provide direct access to the respective singlet GPDs at the particular kinematic situation $x = \xi$.

2.4.2   The Beam Charge and Spin Difference

The beam charge and spin difference of cross sections for an unpolarised target and a polarised lepton beam reads:

$$\mathcal{D}_{CS,U} = \mathrm{d}\sigma^{\overset{+}{\leftarrow}} - \mathrm{d}\sigma^{\overset{-}{\rightarrow}} = 2\Big(|P_\mu|\mathrm{d}\sigma_{\mathrm{pol}}^{\mathrm{DVCS}} + |e_\mu|\mathrm{Re}I\Big),$$

while the beam charge and polarisation are denoted by $+-$ and $\rightarrow \leftarrow$. The remaining two terms are explicitly given by [45]:

$$\mathrm{d}\sigma_{pol}^{\mathrm{DVCS}} = \frac{e^6}{y^2 Q^2}\Big\{s_1^{\mathrm{DVCS}}\sin\phi_{\gamma^*\gamma}\Big\}, \tag{2.29}$$

and

$$\mathrm{Re}I = \frac{e^6}{x_{\mathrm{Bj}}y^3 t \mathcal{P}_1(\phi_{\gamma^*\gamma})\mathcal{P}_2(\phi_{\gamma^*\gamma})} \tag{2.30}$$
$$\Big(c_0^I - c_1^I \cos\phi_{\gamma^*\gamma} + \Big\{c_2^I \cos 2\phi_{\gamma^*\gamma} - c_3^I \cos 3\phi_{\gamma^*\gamma}\Big\}\Big).$$

Kinematically suppressed factors are denoted by {} and the $\phi_{\gamma^*\gamma}$ dependence of the Bethe-Heitler lepton propagators are depicted by $\mathcal{P}_1$ and $\mathcal{P}_2$ according to Ref. [45]. The analysis of the $\phi_{\gamma^*\gamma}$ dependence[8] will thus be most sensitive to the coefficients c_0^I and c_1^I. Neglecting again kinematically suppressed factors[9] within the coefficients c_0^I and c_1^I, one observes that they are mostly sensitive to the real part of the Compton Form Factor $\mathcal{H}$ [45]:

$$c_0^I, c_1^I \propto \mathrm{Re}(F_1 \mathcal{H}),$$

which provides information on the GPD H in the sense of Eqs. 2.27 and 2.28.

2.4.3   The Beam Charge and Spin Sum

The beam charge and spin sum of cross sections for an unpolarised target and a polarised lepton beam reads:

$$\mathcal{S}_{CS,U} = \mathrm{d}\sigma^{\overset{+}{\leftarrow}} + \mathrm{d}\sigma^{\overset{-}{\rightarrow}} = 2\Big(\mathrm{d}\sigma^{BH} + \mathrm{d}\sigma_{\mathrm{unpol}}^{\mathrm{DVCS}} + |e_\mu||P_\mu|\mathrm{Im}I\Big).$$

---

[8]The notation of the coefficients s_i and c_i follows Ref. [45], where the complete expansion of the coefficients can be found. The difference in the defintion of the $\phi_{\gamma^*\gamma}$ angle within this thesis and the ϕ angle within Ref. [45] leads to sign changes in the angular modulations. The angles are related via $\pi - \phi = \phi_{\gamma^*\gamma}$ and this is taken into account within Eqs. 2.29, 2.30 and (2.31).

[9]This refers to terms which are kinematically suppressed with respect to the COMPASS kinematics and not in general suppressed.

The terms $d\sigma^{BH}$, $d\sigma_{unpol}^{DVCS}$ and $\mathrm{Im}I$ are given as follows [45]:

$$d\sigma^{BH} = \frac{e^6}{x_{Bj}y^2(1+\epsilon^2)^2 t \mathcal{P}_1(\phi_{\gamma^*\gamma})\mathcal{P}_2(\phi_{\gamma^*\gamma})}\left(c_0^{BH} - c_1^{BH}\cos\phi_{\gamma^*\gamma} + c_2^{BH}\cos 2\phi_{\gamma^*\gamma}\right),$$

$$d\sigma_{unpol}^{DVCS} = \frac{e^6}{y^2 Q^2}\left(c_0^{DVCS} - \left\{c_1^{DVCS}\cos\phi_{\gamma^*\gamma} - c_2^{DVCS}\cos 2\phi_{\gamma^*\gamma}\right\}\right),$$

$$\mathrm{Im}I = \frac{e^6}{x_{Bj}y^3 t \mathcal{P}_1(\phi_{\gamma^*\gamma})\mathcal{P}_2(\phi_{\gamma^*\gamma})}\left(-s_1^I \sin\phi_{\gamma^*\gamma} + \left\{s_2^I \sin 2\phi_{\gamma^*\gamma}\right\}\right), \qquad (2.31)$$

with ϵ^2 given by:

$$\epsilon^2 = 4x_{Bj}^2 \frac{M^2}{Q^2}.$$

The coefficients marked with the superscript BH are calculable within QED, while the well measured Form Factors F_1 and F_2 are the only experimental input needed. Again kinematically suppressed terms are marked with {}. After a subtraction of the Bethe-Heitler contribution the analysis of the angular $\phi_{\gamma^*\gamma}$ dependence can provide the coefficient s_1^I, which is given in terms of the Compton Form Factors as follows [45]:

$$s_1^I \propto \mathrm{Im}\left(F_1\mathcal{H} + \frac{x_{Bj}}{2-x_{Bj}}(F_1+F_2)\tilde{\mathcal{H}} - \frac{\Delta^2}{4M^2}F_2\mathcal{E}\right) \propto \mathrm{Im}(F_1\mathcal{H}),$$

and one gains sensitivity to the Compton Form Factor $\mathcal{H}_{\mathrm{Im}}$, which is connected to the GPD H in terms of Eq. 2.27 and (2.28).

The extraction of the leading twist-2 quantity c_0^{DVCS} is achieved by the subtraction of the Bethe-Heitler contribution and an integration in $\phi_{\gamma^*\gamma}$, which causes the cancellation of all $\phi_{\gamma^*\gamma}$ dependent terms. The coefficient c_0^{DVCS} reads explicitly in terms of the Compton Form Factors [45]:

$$c_0^{DVCS} = 2(2-y+y^2)\frac{1}{(2-x_{Bj})^2}\left\{4(1-x_{Bj})\left(\mathcal{H}\mathcal{H}^* + \tilde{\mathcal{H}}\tilde{\mathcal{H}}^*\right) - x_{Bj}^2 \frac{t}{4M^2}\tilde{\mathcal{E}}\tilde{\mathcal{E}}^*\right.$$
$$\left. - x_{Bj}^2\left(\mathcal{H}\mathcal{E}^* + \mathcal{E}\mathcal{H}^* + \tilde{\mathcal{H}}\tilde{\mathcal{E}}^* + \tilde{\mathcal{E}}\tilde{\mathcal{H}}^*\right) - \left(x_{Bj}^2 + (2-x_{Bj})^2\frac{t}{4M^2}\right)\mathcal{E}\mathcal{E}^*\right\}.$$
$$(2.32)$$

Neglecting again kinematically suppressed terms and the contribution of $\tilde{\mathcal{H}}$ the coefficient c_0^{DVCS} provides mainly information on the real and imaginary part of the Compton Form Factor $\mathcal{H}$:

$$c_0^{\text{DVCS}} \propto \mathcal{H}_{\text{Re}}^2 + \mathcal{H}_{\text{Im}}^2.$$

The extraction of the t-dependence of the quantity c_0^{DVCS} is the main focus of this thesis.

2.4.4 DVCS in the Valence Quark Region

Recently an application of the nucleon tomography described in Sect. 2.3.4 was performed within Ref. [59]. The combined DVCS observables from HERMES,[10] CLAS[11] and Hall A[12] were used in order to extract simultaneously all eight Compton Form Factors at a given value of ξ and t with a least squares method, incorporating the eight Form Factors as free parameters. Since the experimental observables receive in general contributions from several Compton Form Factors (see e.g. Eq. 2.32) the problem is in principle underconstrained and model dependent limits have to be imposed on the variation of the Compton Form Factors. Imposing these limits in a conservative way and in case an observable is dominated by a certain Compton Form Factor, the Form Factor can be extracted with a finite error bar. Figure 2.14 shows the extracted values of the imaginary part of the Compton Form Factor $\mathcal{H}$ as a function of ξ and $-t$. The Compton Form Factor is denoted by $\mathcal{H}_{Im}$ according to Eq. 2.28.[13]

For each set of $\mathcal{H}_{Im}$ at a certain value of ξ the $|t|$ dependence of $\mathcal{H}_{Im}$ is extracted according to an exponential law:

$$\mathcal{H}_{Im}(\xi, t) \propto e^{B(\xi)t}, \tag{2.33}$$

which leads to the results of $B(\xi)$ shown on the left side of Fig. 2.15. The right side of Fig. 2.15 shows the conversion of $B(\xi)$ into $< b_\perp^2 > (x)$ using the relation

$$< b_\perp^2 > (x) = 4B_0(x) \approx k \, 4B(\xi). \tag{2.34}$$

The correction factor k accounts for the following facts. The extracted quantities are the t-slopes $B(\xi)$ of the imaginary part of the Compton Form Factor $\mathcal{H}$, given according to Eq. 2.28 by the singlet GPD H_+ at $x = \xi$. The quantity $< b_\perp^2 > (x)$ denotes in contrast the mean valence quark radius squared, which is related to $B_0(x)$ the t-dependence of the valence GPD:

---

[10]**HER**A **ME**asurement of **S**pin: Fixed target experiment at DESY's HERA facility to explore the nucleon spin.

[11]**C**EBAF **L**arge **A**cceptance **S**pectrometer: Fixed target experiment located at the experimental Hall B at Jefferson Laboratory.

[12]Fixed target experiment located at the experimental **Hall A** at Jefferson Laboratory.

[13]Note that the authors of Ref. [59] do not absorb the factor of π within the definition of $\mathcal{H}_{Im}$. Thus, the factor π has to be removed in Eq. 2.28 to be in accordance with Ref. [59].

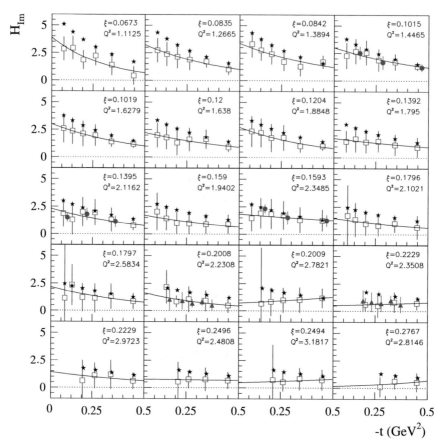

Fig. 2.14 t-dependence of the Compton Form Factor (CFF) $\mathcal{H}_{im}$. Open squares: Results of the CLAS σ and $\Delta\sigma$ fits with eight CFFs as free parameters. Solid circles: results of the fit to CLAS σ and $\Delta\sigma$, as well as longitudinally polarised target and double beam-target polarised asymmetries, with the eight CFFs as free parameters. Solid triangles: results of the Hall A σ and $\Delta\sigma$ fit with the eight CFFs as free parameters. Stars: VGG reference DFFs. The solid curve shows an exponential fit of the open squares according to Eq. 2.33 (see Ref. [59] and references within for the experimental input)

$$H^f_-(x, \xi, t) = H^f(x, \xi, t) + H^f(-x, \xi, t),$$

at $\xi = 0$. The difference between the two slopes B_0 and B is studied in several models and a single correction factor $k = B_0/B = 0.925 \pm 0.025$ is applied by the authors of Ref. [59], in order to convert the left side of Fig. 2.15 into the right side via Eq. 2.34.

The prediction inside the right plot of Fig. 2.15 corresponds to a Regge type ansatz for B_0:

$$B_0(x) = a_{B0} \ln 1/x. \tag{2.35}$$

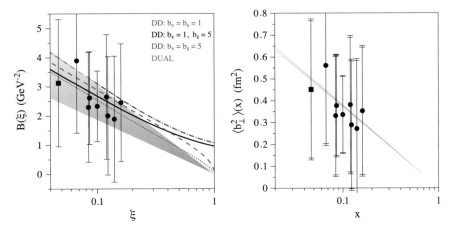

Fig. 2.15 Left: t-slope B of $\mathcal{H}_{Im}$ as a function of ξ. The theory curves correspond with the dual model and the double distribution (DD) model for three choices of the valence (sea) profile parameters b_v (b_s), as indicated. Right: x-dependence of $< b_\perp^2 >$. The band corresponds to the ansatz given by Eqs. 2.35 and 2.36. The data points correspond to the plot on the left side, using Eq. 2.34. The outer error bars take the model uncertainty introduced by the factor k of Eq. 2.34 into account. (see Ref. [59] and references within for the experimental input)

This yields a similar form for the valence GPD H_-^f as discussed in Eq. 2.21 for the GPD H^f, which reads:

$$H_-^f(x, 0, t) = q_v^f(x)e^{a_{B0}\ln 1/x}.$$

Assuming the same x-dependence of B_0 for the up and down quark flavours f, exploiting the connection to the known Form Factor F_1 via Eq. 2.20 by using the valence quark distributions $q_v^f(x)$, the parameter a_{B0} is estimated to [59]:

$$a_{B0} = (1.05 \pm 0.02)\,\text{GeV}^{-2}. \tag{2.36}$$

2.4.5 DVCS in the Region of Sea Quarks and Gluons

The H1[14] and ZEUS[15] experiments at the HERA[16] collider have measured the pure DVCS cross section, which is directly proportional to the contribution c_0^{DVCS}, as described in Sect. 2.4.3. This procedure is feasible as soon as the DVCS process

---

[14]Experiment using the general purpose detector H1 build around one of the ep collision points of HERA.

[15]$ZEY\Sigma$: $Z\eta\tau\eta\sigma\iota\varsigma$ $\kappa\alpha\vartheta'$ $E\nu\rho\epsilon\tau\eta\varsigma$ $\Upsilon\pi\phi\kappa\epsilon\iota\mu\epsilon\nu\eta\varsigma$ $\Sigma\upsilon\mu\mu\epsilon\tau\rho\iota\alpha\varsigma$. Greek for "Search for enlightment related to fundamental symmetries": Experiment around another ep collision point of HERA.

[16]**H**adron **E**lektron **R**ing **A**nlage: Particle accelerator at DESY (**D**eutsches **E**lektron **SY**nchrotron).

Fig. 2.16 DVCS diagram
for two gluon exchange [60]

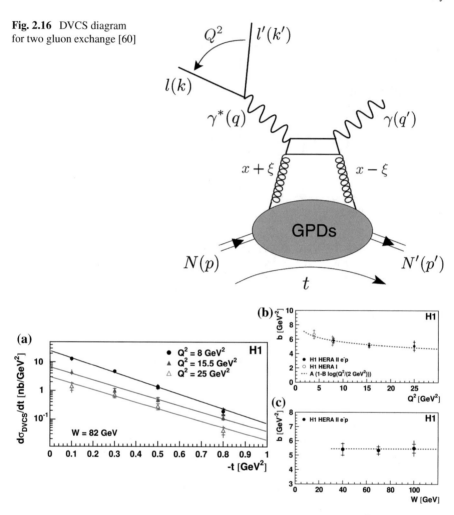

Fig. 2.17 Left: The DVCS cross section, differential in t, for three values of Q^2 expressed at W = 82 GeV/c^2. The curves correspond to a fit of the form $d\sigma \propto e^{Bt}$. Right: The values of B as a function of Q^2 (top) and W (bottom) [63]

becomes dominant with respect to the Bethe-Heitler process, which is the case for the high centre of mass energy achieved at HERA in the collider mode, where an electron beam with a momentum of 27 GeV/c collides with a proton beam of 160 GeV/c. It is also feasible with the high energy muon beam at the COMPASS-II experiment, as it will be demonstrated throughout this thesis. The range in x_{Bj} covered by H1 and ZEUS goes from 10^{-4} to 10^{-2}. At such small values of x_{Bj} the gluon exchange, shown in Fig. 2.16, plays also an important role in addition to the leading order process of the quark photon interaction, shown in Figs. 2.9 and 2.12.

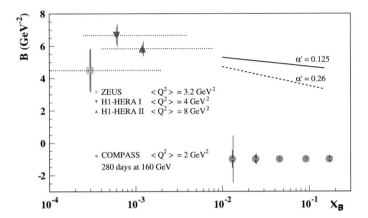

Fig. 2.18 Blue and Green: The parameter B of the DVCS cross section, according to Eq. 2.37, for the three lowest bins in Q^2, as measured by H1 and Zeus [64–66]. Red: COMPASS-II projections for measuring the x_{Bj} dependence of the t-slope parameter $B(x_{Bj})$ of the DVCS cross section, calculated for $1\,(\text{GeV/c})^2 < Q^2 < 8\,(\text{GeV/c})^2$. The left vertical error bar on each data point indicates the statistical error only, while the right one includes also the quadratically added systematic uncertainty [43]. The black lines correspond to an ansatz of the form $B(x_{Bj}) = B_0 + 2\alpha' \ln\left(\frac{x_0}{x_{Bj}}\right)$ while the value for $\alpha' = 0.26\,(\text{GeV/c})^{-2}$ is inspired by the value obtained for Pomeron exchange in soft scattering processes [61]

The t-dependence of the DVCS cross section measured at H1 and ZEUS was found to be in agreement with a Regge behaviour of the form:

$$\frac{d\sigma}{dt} \propto e^{Bt}, \tag{2.37}$$

as shown on the left side of Fig. 2.17. For exclusive meson production it is observed that B is dependent on W [61, 62]. For decreasing W the value of B decreases, which means that the size of the scattering object becomes smaller. This so called shrinkage effect was not observed for DVCS, as it can be seen within the bottom right plot of Fig. 2.17. In case of DVCS the parameter B shows a weak dependence on Q^2 though, as illustrated on the top right side of Fig. 2.17.

The values of the slope parameter B measured at H1 and ZEUS for the lowest Q^2 bins are summarised within Fig. 2.18 and shall be confronted with the findings at COMPASS-II in the x_{Bj} region of 10^{-2} to 0.2, within the DVCS pilot run (Sect. 7.7) and with the future results of the dedicated DVCS data taking in 2016 and 2017.

References

1. R. Frisch, O. Stern , Über die magnetische Ablenkung von Wasserstoffmolekülen und das magnetische Moment des Protons. I. Zeitschrift für Physik **85**, 4–16 (1933). https://doi.org/10.1007/BF01330773
2. M.N. Rosenbluth, High energy elastic scattering of electrons on protons. Phys. Rev. **79**, 615–619 (1950). https://doi.org/10.1103/PhysRev.79.615
3. R.W. McAllister, R. Hofstadter, Elastic scattering of 188 MeV electrons from the proton and the alpha particle. Phys. Rev. **102**, 851–856 (1956). https://doi.org/10.1103/PhysRev.102.851
4. R. Hofstadter, Electron scattering and nuclear structure. Rev. Mod. Phys. **28**, 214–254 (1956). https://doi.org/10.1103/RevModPhys.28.214
5. C.F. Perdrisat, V. Punjabi, M. Vanderhaeghen, Nucleon electromagnetic form factors. Prog. Part. Nucl. Phys. **59**, 694–764 (2007). https://doi.org/10.1016/j.ppnp.2007.05.001
6. L.L. Foldy, The electromagnetic properties of Dirac particles. Phys. Rev. **87**, 688–693 (1952). https://doi.org/10.1103/PhysRev.87.688
7. L.N. Hand, D.G. Miller, R. Wilson, Electric and magnetic form factors of the nucleon. Rev. Mod. Phys. **35**, 335–349 (1963). https://doi.org/10.1103/RevModPhys.35.335
8. A1 Collaboration, J.C. Bernauer et al., Electric and magnetic form factors of the proton. Phys. Rev. C **90**, 015206 (2014). https://doi.org/10.1103/PhysRevC.90.015206
9. R. Pohl et al., The size of the proton. Nature **466**, 213–216 (2010). https://doi.org/10.1038/nature09250
10. P.J. Mohr, D.B. Newell, B.N. Taylor, CODATA recommended values of the fundamental physical constants: 2014*. Rev. Mod. Phys. **88**, 035009 (2016). https://doi.org/10.1103/RevModPhys.88.035009
11. R. Pohl et al., Laser spectroscopy of Muonic deuterium. Science **353**, 669–673 (2016). https://doi.org/10.1126/science.aaf2468
12. A. Manohar, *An Introduction to Spin Dependent Deep Inelastic Scattering*. Lake Louise Winter Institute: Symmetry and Spin in the Standard Model Lake Louise, Alberta, Canada, 23–29 Feb 1992, 1–46 (1992), arXiv:hep-ph/9204208
13. C. Patrignani et al., Review of particle physics. Chin. Phys. C **40**, 100001 (2016). https://doi.org/10.1088/1674-1137/40/10/100001
14. V. Barone, P.G. Ratcliffe, *Transverse Spin Physics* (World Scientific, Singapore, 2003)
15. M. Anselmino, A. Efremov, E. Leader, The theory and phenomenology of polarized deep inelastic scattering. Phys. Rep. **261**, 1–124 (1995). https://doi.org/10.1016/0370-1573(95)00011-5, [Erratum: Phys. Rep. **281**, 399(1997)]
16. C.G. Callan, D.J. Gross, High-energy electroproduction and the constitution of the electric current. Phys. Rev. Lett. **22**, 156–159 (1969). https://doi.org/10.1103/PhysRevLett.22.156
17. V.N. Gribov, L.N. Lipatov, Deep inelastic e p scattering in perturbation theory. Sov. J. Nucl. Phys. **15**, 438–450 (1972), [Yad. Fiz.15,781(1972)]
18. L.N. Lipatov, The parton model and perturbation theory. Sov. J. Nucl. Phys. **20**, 94–102 (1975), [Yad. Fiz.20,181(1974)]
19. G. Altarelli, G. Parisi, Asymptotic freedom in Parton language. Nucl. Phys. B **126**, 298–318 (1977). https://doi.org/10.1016/0550-3213(77)90384-4
20. Y.L. Dokshitzer, Calculation of the structure functions for deep inelastic scattering and e+ e-annihilation by perturbation theory in quantum chromodynamics. Sov. Phys. JETP **46**, 641–653 (1977), [Zh. Eksp. Teor. Fiz.73,1216(1977)]
21. V.S. Fadin, E.A. Kuraev, L.N. Lipatov, On the pomeranchuk singularity in asymptotically free theories. Phys. Lett. B **60**, 50–52 (1975). https://doi.org/10.1016/0370-2693(75)90524-9
22. V.S. Fadin, E.A. Kuraev, L.N. Lipatov, Multi-reggeon processes in the Yang-Mills Theory. Sov. Phys. JETP **44**, 443–450 (1976), [Zh. Eksp. Teor. Fiz.71,840(1976)]

23. V.S. Fadin, E.A. Kuraev, L.N. Lipatov, The Pomeranchuk singularity in nonabelian gauge theories. Sov. Phys. JETP **45**, 199–204 (1977), [Zh. Eksp. Teor. Fiz.72,377(1977)]
24. I.I. Balitsky, L.N. Lipatov, The Pomeranchuk singularity in quantum chromodynamics. Sov. J. Nucl. Phys. **28**, 822–829 (1978), [Yad. Fiz.28,1597(1978)]
25. B. Lampe, E. Reya, Spin physics and polarized structure functions. Phys. Rep. **332**, 1–163 (2000). https://doi.org/10.1016/S0370-1573(99)00100-3
26. S.A. Larin, The next-to-leading QCD approximation to the Ellis-Jaffe sum rule. Phys. Lett. B **334**, 192–198 (1994). https://doi.org/10.1016/0370-2693(94)90610-6
27. J. Kodaira, QCD higher order effects in polarized electroproduction: flavor singlet coefficient functions. Nucl. Phys. B **165**, 129–140 (1980). https://doi.org/10.1016/0550-3213(80)90310-7
28. EMC Collaboration, J. Ashman et al., A measurement of the spin asymmetry and determination of the structure function g(1) in deep inelastic muon-proton scattering. Phys. Lett. **B206**, 364 (1988). https://doi.org/10.1016/0370-2693(88)91523-7
29. J.R. Ellis, R.L. Jaffe, A sum rule for deep inelastic electroproduction from polarized protons. Phys. Rev. D **9**, 1444 (1974). https://doi.org/10.1103/PhysRevD.9.1444, [Erratum: Phys. Rev. D **10**, 1669 (1974)]
30. COMPASS Collaboration, C. Adolph et al., The spin structure function g_1^p of the proton and a test of the Bjorken sum rule. Phys. Lett. B **753**, 18–28 (2016). https://doi.org/10.1016/j.physletb.2015.11.064
31. COMPASS Collaboration, C. Adolph et al., Final COMPASS results on the deuteron spin-dependent structure function g_1^d and the Bjorken sum rule. (2016), arXiv:1612.00620
32. COMPASS Collaboration, M.G. Alekseev et al., Quark helicity distributions from longitudinal spin asymmetries in muon–proton and muon–deuteron scattering. Phys. Lett. B **693**, 227–235 (2010). https://doi.org/10.1016/j.physletb.2010.08.034
33. COMPASS Collaboration, C. Adolph et al., Multiplicities of charged kaons from deep-inelastic muon scattering off an isoscalar target. Phys. Lett. B **767**, 133–141 (2017). https://doi.org/10.1016/j.physletb.2017.01.053
34. D. de Florian et al., Global analysis of helicity parton densities and their uncertainties. Phys. Rev. Lett. **101**, 072001 (2008). https://doi.org/10.1103/PhysRevLett.101.072001
35. D. de Florian et al., Extraction of spin-dependent parton densities and their uncertainties. Phys. Rev. D **80**, 034030 (2009). https://doi.org/10.1103/PhysRevD.80.034030
36. COMPASS Collaboration, C. Adolph et al., Leading and next-to-leading order gluon polarization in the nucleon and longitudinal double spin asymmetries from open charm muoproduction. Phys. Rev. D **87**, 052018 (2013). https://doi.org/10.1103/PhysRevD.87.052018
37. COMPASS Collaboration, C. Adolph et al., Leading order determination of the gluon polarisation from DIS events with high-p_T hadron pairs. Phys. Lett. B **718**, 922–930 (2013). https://doi.org/10.1016/j.physletb.2012.11.056
38. A. Bravar, K. Kurek, R. Windmolders, POLDIS A Monte Carlo for POLarized (semi-inclusive) Deep Inelastic Scattering. Comput. Phys. Commun. **105**, 42–61 (1997). https://doi.org/10.1016/S0010-4655(97)00063-5
39. COMPASS Collaboration, C. Adolph et al., Leading-order determination of the gluon polarisation using a novel method. (2015), arXiv:1512.05053
40. D. de Florian et al., Evidence for polarization of gluons in the proton. Phys. Rev. Lett. **113**, 012001 (2014). https://doi.org/10.1103/PhysRevLett.113.012001
41. R.L. Jaffe, A. Manohar, The g1 problem: deep inelastic electron scattering and the spin of the proton. Nucl. Phys. B **337**, 509–546 (1990). https://doi.org/10.1016/0550-3213(90)90506-9
42. X. Ji, Gauge-invariant decomposition of nucleon spin. Phys. Rev. Lett. **78**, 610–613 (1997). https://doi.org/10.1103/PhysRevLett.78.610

43. COMPASS Collaboration, F. Gautheron et al., COMPASS-II Proposal. CERN-SPSC-2010-014, SPSC-P-340, (2010), http://wwwcompass.cern.ch/compass/proposal/compass-II_proposal/compass-II_proposal.pdf

44. J.C. Collins, A. Freund, Proof of factorization for deeply virtual Compton scattering in QCD. Phys. Rev. D **59**, 074009 (1999). https://doi.org/10.1103/PhysRevD.59.074009

45. A.V. Belitsky, D. Mueller, A. Kirchner, Theory of deeply virtual Compton scattering on the nucleon. Nucl. Phys. B **629**, 323–392 (2002). https://doi.org/10.1016/S0550-3213(02)00144-X

46. M. Diehl, Generalized Parton distributions. Phys. Rep. **388**, 41–277 (2003). https://doi.org/10.1016/j.physrep.2003.08.002

47. A.V. Efremov, A.V. Radyushkin, Factorization and asymptotic behaviour of Pion form factor in QCD. Phys. Lett. B **94**, 245–250 (1980). https://doi.org/10.1016/0370-2693(80)90869-2

48. G.P. Lepage, S.J. Brodsky, Exclusive processes in quantum chromodynamics: evolution equations for hadronic wavefunctions and the form factors of mesons. Phys. Lett. B **87**, 359–365 (1979). https://doi.org/10.1016/0370-2693(79)90554-9

49. D. Müller et al., Wave functions, evolution equations and evolution kernels from light ray operators of QCD. Fortschr. Phys. **42**, 101–141 (1994). https://doi.org/10.1002/prop.2190420202

50. A.V. Radyushkin, Scaling limit of deeply virtual compton scattering. Phys. Lett. B **380**, 417–425 (1996). https://doi.org/10.1016/0370-2693(96)00528-X

51. A.V. Radyushkin, Nonforward Parton distributions. Phys. Rev. D **56**, 5524–5557 (1997). https://doi.org/10.1103/PhysRevD.56.5524

52. M.V. Polyakov, C. Weiss, Skewed and double distributions in the Pion and the nucleon. Phys. Rev. D **60**, 114017 (1999). https://doi.org/10.1103/PhysRevD.60.114017

53. M. Burkardt, Impact parameter space interpretation for generalized parton distributions. Int. J. Mod. Phys. A **18**, 173–208 (2003). https://doi.org/10.1142/S0217751X03012370

54. M. Burkardt, C.A. Miller, W.-D. Nowak, Spin-polarized high-energy scattering of charged leptons on nucleons. Rep. Prog. Phys. **73**, 016201 (2010). https://doi.org/10.1088/0034-4885/73/1/016201

55. M. Diehl, Generalized Parton distributions in impact parameter space. Eur. Phys. J. C **25**, 223–232 (2002). https://doi.org/10.1007/s10052-002-1016-9, [Erratum: Eur. Phys. J. C **31**, 277 (2003)]

56. M. Burkardt, GPDs with $\zeta \neq 0$. (2007), arXiv:0711.1881

57. S.J. Brodsky, M. Diehl, D.S. Hwang, Light cone wave function representation of deeply virtual Compton scattering. Nucl. Phys. B **596**, 99–124 (2001). https://doi.org/10.1016/S0550-3213(00)00695-7

58. J. Kiefer, First measurement of the transverse-target single-spin asymmetry in exclusive muon-production of Rho Mesons at COMPASS. Diploma thesis, Albert Ludwigs Universität Freiburg (2007)

59. R. Dupre, M. Guidal, M. Vanderhaeghen, Tomographic image of the proton. Phys. Rev. D **95**, 011501 (2017). https://doi.org/10.1103/PhysRevD.95.011501

60. N. d'Hose, S. Niccolai, A. Rostomyan, Experimental overview of deeply virtual compton scattering. Eur. Phys. J. A **52**, 151 (2016). https://doi.org/10.1140/epja/i2016-16151-9

61. ZEUS Collaboration, S. Chekanov et al., Exclusive electroproduction of J/Ψ mesons at HERA. Nucl. Phys. B **695**, 3–37 (2004). https://doi.org/10.1016/j.nuclphysb.2004.06.034

62. H1 Collaboration, F.D. Aaron et al., Diffractive electroproduction of ρ and ϕ Mesons at HERA. JHEP **05**, 032 (2010). https://doi.org/10.1007/JHEP05(2010)032

63. H1 Collaboration, F.D. Aaron et al., Measurement of deeply virtual Compton scattering and its t-dependence at HERA. Phys. Lett. B **659**, 796–806 (2008). https://doi.org/10.1016/j.physletb.2007.11.093

64. H1 Collaboration, F.D. Aaron et al., Deeply virtual Compton scattering and its beam charge asymmetry in $e^{\pm} p$ collisions at HERA. Phys. Lett. B **681**, 391–399 (2009). https://doi.org/10.1016/j.physletb.2009.10.035
65. H1 Collaboration, A. Aktas et al., Measurement of deeply virtual compton scattering at HERA. Eur. Phys. J. C **44**, 1–11 (2005). https://doi.org/10.1140/epjc/s2005-02345-3
66. ZEUS Collaboration, S. Chekanov et al., A measurement of the Q^2, W and t dependences of deeply virtual Compton scattering at HERA. JHEP **2009**, 108 (2009). https://doi.org/10.1088/1126-6708/2009/05/108

Chapter 3
The COMPASS-II Experiment

The COMPASS-II experiment is a fixed target experiment, located at the CERN Prevessin area at the end of the M2-beamline of the Super Proton Synchrotron. The scattering of high energy leptons or hadrons of a nucleon target allows studying the spin structure of the nucleon and performing hadron spectroscopy. This chapter is supposed to give an overview of the COMPASS-II experimental setup, while the focus is put on the experimental tools necessary to investigate the spin structure of the nucleon. The major upgrades of COMPASS, dedicated to the Deeply Virtual Compton Scattering measurement, are a third electromagnetic calorimeter, ECal0, and the proton recoil detector CAMERA.

3.1 The Beam

The COMPASS-II experiment can switch easily between electron, muon and hadron beams. The beam itself is generated by proton collisions with a beryllium target (T6) of variable length, to adjust for different intensities of the secondary, respectively teriary beams.

Apart from the energy scale and luminosity the production of the tertiary muon beam shows quite some similarity with the mechanisms involved in the creation of cosmic muons. Protons with an energy of up to 450 GeV, extracted from the Super Proton Synchrotron, are scattered of a beryllium material block. The parity violating decay of the pions and kaons, produced in this collision, into μ^+ and ν (respectively μ^- and $\bar{\nu}$) allows for a polarised muon beam. The beam polarisation is dependent on the ratio of the meson and muon momenta. Thus, in order to reach a polarisation of $(80 \pm 5)\%$, it is mandatory to perform a momentum selection. The momentum selection is achieved by bending magnets within a 600 m long tunnel. The fraction of hadrons which did not decay is filtered by a second hadron absorber. Within an 800 m long tunnel the beam is then injected into the COMPASS-II experiment. The spill structure, which consists of an on- and off-spill phase, may vary depending on other

© Springer International Publishing AG, part of Springer Nature 2018
P. Jörg, *Exploring the Size of the Proton*, Springer Theses,
https://doi.org/10.1007/978-3-319-90290-6_3

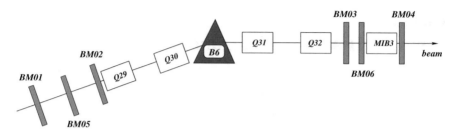

Fig. 3.1 Positioning of the Beam Momentum Stations (BMS) [1]

consumers of the Super Proton Synchrotron. During the 2012 DVCS measurement the on-spill phase was set to 9.6 s[1] with an off-spill phase of 38.4 s.

Since the beam particles are created in decays, an overall beam momentum spread of approximately five percent is tolerated. However, to guarantee a precise measurement of the beam momentum, a momentum measurement of each individual beam particle is performed. This is particularly important since the beam polarisation depends on the muon momentum. The momentum measurement is realised by the so called Beam Momentum Stations (BMS). As shown in Fig. 3.1, by placing a bending magnet (B6) between two scintillating fiber detectors (BM05, BM06) and four hodoscopes (BM01-BM04), the momentum of the beam particle is measured from the radius of curvature of the particle in the magnetic field of the bending magnet with an uncertainty of about one percent.

The beam, which is accompanied by a so called halo, is then focused on the target.

3.2 The Target

In the framework of the muon programme polarised NH_3 and LID targets were used in the past. In 2012 for the first time a detection of the recoiled target proton was realised. An unpolarised LH_2 target, surrounded by the proton recoil detector CAMERA, was installed in the target region. A schematic drawing of the target is shown in Fig. 3.2. To achieve a luminostiy of $10^{33} \frac{1}{cm^2 s}$ with the anti-muon beam, a target length of 2.5 m was chosen. While the target is coverd in detail in Ref. [2], it is worth to outline the two major technical challenges of the target construction:

- A minimum amount of material: From the "physics" point of view there is a particular interest in small momentum transfer to the target proton. In order to measure the recoiled target proton down to a momentum of 260 MeV/c, its absorption

---

[1]This value corresponds to the amount of time during which a beam hits the COMPASS-II target. Technically, the begin of spill signal arrives 1 s in advance of the beam. Furthermore, to guarantee a good beam quality within the analysis, a window between 1 and 10.4 s with respect to the begin of spill signal is used later. This window is 0.2 s shorter than the value given above.

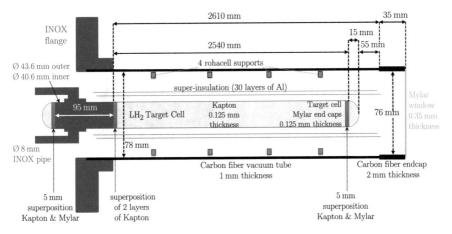

Fig. 3.2 A schematic side view of the target cell and vacuum chamber (picture adapted from Ref. [2] by Ref. [3])

within the target material has to be avoided. Hence, the target cryostat material had to be minimised drastically.

- Homogeneous LH$_2$ density: To achieve a precise measurement of the luminosity, a homogeneous density of the LH$_2$ inside the target volume with a minimal gas phase has to be realised. Together with a precise knowledge of the muon flux a measurement of the luminosity within an uncertainty of a few percent will then be realised.

3.3 The Spectrometer

The COMPASS-II spectrometer is shown in Fig. 3.3. It extends over a length of approximately 60 m and measures in a sophisticated manner the mass, energy and momentum of elementary particles. For a full description of a particles four-momentum it is of course sufficient to measure only two of the three properties, since they are correlated by the relation $p_\nu p^\nu = m^2 c^2$. The quantity p_ν indicates the ν component of the particle four-momentum, m the particle mass and c the speed of light. The spectrometer is divided into two stages. Each stage comprises one of the two spectrometer dipole magnets (SM1, SM2). The particle momentum can be determined from the radius of curvature of the particle trajectory together with the precise knowledge of the magnetic field, interacting with the particle at each space point. The first spectrometer stage (LAS[2]) is designed for large scattering angles up

---

[2]Large Angle Spectrometer.

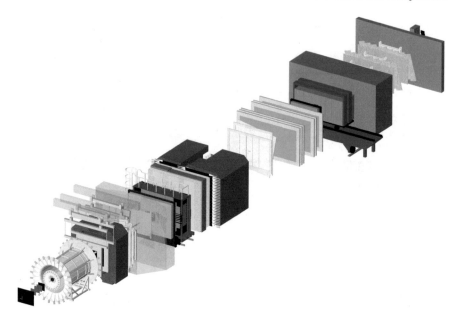

Fig. 3.3 The COMPASS-II spectrometer [3]

to 180 mrad and is located close to the interaction vertex. The second spectrometer stage (SAS[3]) allows for a scattering angle acceptance below 30 mrad.

3.3.1 Track Reconstruction

For the track reconstruction the use of the appropriate detector technology depends on the distance to the beam and the interaction vertex. Close to the beam axis, a high rate stability as well as high time and spatial detector resolutions are necessary. At a larger distance to the beam axis the requirements on rate stability and resolution can be relaxed, while putting the focus on large area coverage. Table 3.1 shows the spatial coverage and resolution of the different detector types used in the COMPASS-II experiment. A detailed description of the different detector technologies can be found in Ref. [1].

---

[3]Small Angle Spectrometer.

Table 3.1 Overview of the different track reconstruction detectors at the COMPASS-II experiment. The quantity A denotes the active detector area, δ_x and δ_t the achievable spatial and time resolution of the detectors [4]

Class	Type	A/cm^2	δ_x/μm	δ_t/ns
VSAT	SCIFI[a]	3.9^2–12.3^2	130–210	0.4
	SILICON detectors	5×7	8–11	2.5
	Pixel-GEM[b]	10×10	95	9.9
SAT	GEM[b]	31×31	70	12
	MicroMegas[c]	40×40	90	9
LAT	MWPC [d]	$178 \times (90–120)$	1600	
	DC[e]	180×127	190–500	
	Straw[f]	280×323	190	

[a]**SCI**ntillating **FI**bers
[b]**G**as **E**lectron **M**ultiplier
[c]**Micro-Me**sh **Ga**seous **S**tructure
[d]**M**ulti**Wi**re **P**roportional **C**hambers
[e]**D**rift **C**hambers
[f]due to the visual similarity between kapton tubes and straws

3.4 Particle Identification

The particle identification can be seperated into three parts: the muon filters, the calorimeters and the RICH detector.[4]

- **Muon Filters**: The principle for the identification of muons with a momentum of 160 GeV/c relies on the comparably large lifetime of the muon, its low energy loss in matter due to electromagnetic effects and the fact that it is not a strongly interacting particle. The identification is achieved by so called muon walls. The muon walls are absorbers only the muon can pass. Muon walls, accompanied by tracking detectors before and after, are placed in both spectrometer stages.
  At the end of the LAS spectrometer stage the Muon Wall I is placed. It is build of a 60 cm thick iron absorber with four drift chamber planes placed before and after the absorber. Particles with a small scattering angle can pass the absorber through a hole in the centre and can thus reach the SAS spectrometer stage.
  The Muon Wall II, which is a 2.4 m concrete material block is placed in the SAS spectrometer stage. A particle is identified as a muon in case its track parameters, given by the corresponding tracking detectors before and after the muon filter, are compatible.
- **Calorimeters**: An electromagnetic calorimeter (ECal1,2) as well as a hadronic calorimeter (HCal1,2) is placed in each spectrometer stage. In 2012, dedicated to the DVCS measurement, a third electromagnetic calorimeter (ECal0) was installed

---

[4]**R**ing **I**maging **CH**erenkov.

right after the target and in front of the RICH detector. It was build to improve the acceptance for photons leaving the target under large polar angles.

The electromagnetic calorimeters are mainly made of lead glass or so called shashlik modules. In case of the lead glass modules the incoming photon produces showers of e^+e^- pairs within the lead glass. The emmited Cherenkov light is then detected by photomultiplier tubes. The intensity of the photomultiplier signals is proportional to the energy deposit of the photon. A shashlik module consists of a stack of alternating layors of lead and scintillating material. The e^+e^- pairs produced in the lead layors radiate visible light within the layors of scintillator material. The visible light is detected by Micro-pixel avalanche photodiodes (MAPD). The energy deposit of the incoming photon is proportional to the collected scintillation light in the various scintillator slices. ECal0 and the centre of ECal2 consist solely of shashlik modules. With both the lead and the shashlik module techniques one can determine 99% of the initial photon energy.

The hadronic calorimeters show quiet some simularity to the shashlik design. They are also so called sampling calorimeters. Build out of alternating layers of iron and scintillator material, the hadronic calorimeters detect the incoming hadron by measuring the showers created in the iron layers within the scintillator material. The relative energy resolution of the electromagnetic calorimeters is ten times better than the resolution of the hadronic calorimeters.

- **RICH**: The RICH detector is a Ring Imaging Cherenkov detector. The angle of the light cone of the Cherenkov light emitted by a charged particle within the RICH gas volume is related to the particle velocity. A measurement of the angle thus allows for a measurement of the particle velocity. In combination with a preceding momentum measurement a particle identification is achieved. A detailed description of the RICH detector can be found within Ref. [5].

  In Ref. [6] the application of the RICH detector for an identification of the outgoing muon was studied. The results suggest that a reasonable identification probability of the muon can only be achieved for a muon momentum below approximately 10 GeV/c, which is not within the DVCS analysis range (see Sect. 7.4). Thus, for an analysis of the DVCS process the RICH can not be used.

3.5 The CAMERA Detector

The CAMERA detector is dedicated to measure the momentum of the slowly recoiled proton in exclusive processes. A photography of the detector is shown in Fig. 3.4. It is build out of two concentric rings of scintillators. Each of the rings itself consists of 24 elements, which are placed concentrically around the liquid hydrogen target. The inner ring will be denoted as ring A, while the outer ring will be denoted as ring B. Each element detects particles inside an azimuthal interval of 15°. To increase the azimuthal resolution, the elements of ring A are displaced by 7.5° with respect to the elements of ring B. A schematic front view of the detector is given in Fig. 3.7. The properties of ring A and B are as follows:

Fig. 3.4 Picture of the CAMERA detector looking into the beam direction. The liquid hydrogen target is placed into the centre, while the carbon tube is removed. The short light guides and the photomultipliers of the inner ring at the upstream end are visible. The scintillators and light guides of the outer ring are visible [7]

- **Ring A**: The scintillating material elements are build out of BC408. The dimensions of each of the elements are $(275 \times 6.3 \times 0.4)\,cm^3$. The scintillators are connected to a approximately 107 cm long light guide at the downstream end and to a 54 cm short light guide at the upstream end. The light guides are connected to photomultiplier tubes of type ET9813B, which have a photocathode of 51 mm diameter. The long light guide is bend by an angle of 45° to improve the photon acceptance, while the short one possesses an angle of 90°.

- **Ring B**: The scintillating elements are build out of BC408 with the dimensions beeing $(360 \times 30 \times 5)\,cm^3$. The geometrical properties of the "fishtail" shaped light guides at both ends are equal. The light guides have a length of approximately 59 cm and are bend by an angle of 90°. They are connected with two photomultiplier tubes of type ET9823B, providing an active area with a diameter of 130 mm.

The scintillating material BC408, used for both types of scintillators, emits its main amount of light with a wavelength of 430 nm. The maximum quantum efficency of the photomultiplier tubes lies between 350 nm and 450 nm.

It may seem striking that the thickness of the ring A elements is much smaller than in case of ring B. The reason is, that in order to measure the proton trajectory, it is necessary to observe a signal in ring B. Thus, the proton has to pass ring A and

Fig. 3.5 Schematic side
view of two corresponding
ring A and B scintillators of
the CAMERA detector

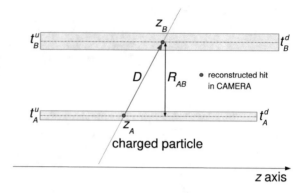

charged particle

z axis

must not be stopped within. This becomes especially critical in case of small values
of $|t|$. The quantity t denotes the square of the four momentum transfer to the proton.
However, a thin ring A decreases the amount of scintillating light and thus reduces
the time resolution dramatically. In 2012 it was aimed for a certain tradeoff, which
allowed a measurement of the proton momentum down to a four-momentum transfer
squared $|t|_{min} = 0.06\,(\text{GeV/c})^2$ with a time resolution of the ring A elements at the
order of 300 ps. Due to bad material quality of the ring A elements, this goal could
not be reached. The degraded time resolution of ring A at the order of up to 400 ps,
was one of the reasons for its exchange before the two years of data taking in 2016
and 2017. The exchange of ring A will shortly be covered in Chap. 9.

Figure 3.5 illustrates the measurement principle of the CAMERA detector. The
trajectory of the recoiling proton, leaving the target, intersects ring A and B. At the
points of intersection a spherical wave of scintillating light is created. It propagates
through the scintillator, being reflected at the horizontal scintillator surfaces, until
it reaches the vertical end points. At the vertical end points it is transported by
light guides to the photomultipliers, where it is converted into a current pulse. The
analogue signal of this current pulse is transmitted to the readout electronics. The
detector readout is performed by the GANDALF[5] Framework, comprising pipelined
sampling ADCs,[6] which convert the analogue photomultiplier signals into digital
signals. A time-stamp and the maximum amplitude information of each of these
digitised photomultiplier signals is extracted inside in total 12 GANDALF modules
and transferred to the data acquisition system.

Apart from overall calibration constants k_A^z and k_B^z, with respect to the COMPASS
coordinate system, the z-positions z_A and z_B of the intersection points are given by
half the difference of the up- and downstream time-stamps times the effective speed
of light $c_{A;B}$ within the corresponding element. Denoting the time-stamp itself with
t, using (u, d) for the up- respectively downstream photomultiplier and (A, B) for
the scintillator type, this can be summarised in Eq. (3.1).

---

[5]Generic **A**dvanced **N**umerical **D**evice for **A**nalog and **L**ogic **F**unctions. For a dense description
and the related references see Sect. 9.2.1.

[6]**A**nalogue to **D**igital **C**onverter.

$$z_A = \frac{1}{2} c_A (t_A^u - t_A^d) + k_A^z,$$

$$z_B = \frac{1}{2} c_B (t_B^u - t_B^d) + k_B^z. \tag{3.1}$$

Using z_A and z_B the distance of flight D is calculated as follows:

$$D = \sqrt{(z_B - z_A)^2 + R_{AB}^2}, \tag{3.2}$$

while R_{AB} denotes the shortest distance between ring A and B. The time of flight T is given by the difference of the mean time values between corresponding ring B and A elements, taking the offset k_T into account:

$$T = \frac{t_B^u + t_B^d}{2} - \frac{t_A^u + t_A^d}{2} + k_T. \tag{3.3}$$

The velocity of the proton between ring A and B in units of the speed of light c follows as:

$$\beta_{AB} = D/T. \tag{3.4}$$

Applying the relations $\beta = p/E$ and $\gamma = E/M$ the momentum of the proton between ring A and B follows as:

$$p_{AB} = m_p \, \beta_{AB} \, \gamma_{AB} = m_p \frac{\beta_{AB}}{\sqrt{1 - \beta_{AB}^2}}, \tag{3.5}$$

while m_p denotes the mass of the proton.

In order to combine the momentum p_{AB} with the COMPASS-II spectrometer measurement of other particles involved in a certain exclusive reaction, the momentum has to be translated into a momentum at the interaction vertex. Hence, one has to take into account energy loss effects inside the material traversed during the flight along the given trajectory. The determination of the calibration constants k_A^z, k_B^z, c_A, c_B and k_T as well as the momentum determination will be part of Sect. 5.1.

3.6 The Trigger System

For the 2012 DVCS data taking period it is convenient to divide the trigger decision into three categories: The muon trigger, the proton trigger and the random trigger.

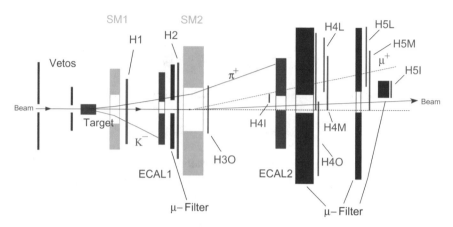

Fig. 3.6 Placement of the trigger hodoscopes for the creation of the muon trigger [8]

3.6.1 The Muon Trigger

In order to identify scattered muons inside a large region of x_{Bj} and Q^2, criteria for the creation of the trigger signal have to be formulated and technically realised. This is achieved with a system of hodoscopes, providing in total a large angular coverage. The common idea is to place one hodoscope upstream and a second one downstream of a muon filter. With this technique it is ensured that the time and space like coincidence at both hodoscopes are related to a muon rather than to a secondary particle or to random noise. The further differentiation of the trigger decision then mainly relies on two methods:

- **Horizontal Target Extrapolation**: Hodoscopes, which are placed horizontally in the x-direction, detect the y-coordinates of the scattered muon at two different z positions. This leads to a determination of the scattering angle between the x-z plane, which is perpendicular to the magnetic field direction, and the plane of the muon trajectory. Using this scattering angle an extrapolation to the y coordinates of the trajectory at the end points of the target can be performed. This method will only work for large scattering angles
- **Vertical Target Extrapolation**: This method relies on the fact that muons scattered inside the target have lost a part of their initial energy. The radius of curvature of their trajectory along the x-z plane will thus be larger compared to an unscattered muon. The combination of two vertically displaced hodoscopes uses this fact for a trigger decision.

Figure 3.6 shows the positions of the different hodoscopes along the COMPASS-II spectrometer. Five different types of triggers can be distinguished.

- **Inner Trigger**: For very small scattering angles vertical target extrapolation together with the hodoscopes H4I and H5I creates the Inner Trigger.

- **Ladder Trigger**: The Ladder Trigger is supposed to detect muons with small scattering angles, large energy deposit and high values of Q^2. The hodoscopes H4L and H5L together with the horizontal target extrapolation method are used.
- **LAS Trigger**: Providing sensitivity for values of Q^2 up to $20\,(\text{GeV/c})^2$, the LAS Trigger uses the hodoscopes H1 and H2.
- **Middle Trigger**: For small scattering angles a combination of vertical and horizontal target extrapolation, using the hodoscopes H4M and H5M, the Middle Trigger is sensitive to the relative energy transfer y in the region $0.1 < y < 0.7$.
- **Outer Trigger**: The Outer Trigger, consisting of the hodoscopes H3O and H4O, covers the full range of the relative energy transfer y and a four-momentum transfer Q^2 of up to $10\,(\text{GeV/c})^2$.

Apart from these types of triggers a veto system located upstream of the target filters the final trigger signal for trigger attempts created by halo particles. A detailed description of the muon trigger system is given in Ref. [8].

3.6.2 The Proton Trigger

The first stage of the proton trigger is shown in Fig. 3.7. An interaction inside one of the ring A elements is combined with all possible interactions within the two corresponding B elements. The combinations are filtered for interactions, which correspond to a longitudinal position within the physical boundaries of the three scintillators. If the time of flight associated to the track lies within $-5\,$ns and $40\,$ns, the first trigger stage is passed. Figure A.61 shows the signature of recoiled protons, detected within elastic pion proton scattering, using the first stage of the proton trigger. The main challenges for the creation of the proton trigger signal are:

- A continous calculation of the time-stamps and amplitudes of each of the 96 photomultiplier signals of the detector [4].
- A high-speed data transfer from the frontend electronics to the trigger electronics [9].
- The generation of the final trigger signal by processing the information of the 96 detector channels on a single module [10].

For the 2017 DVCS measurement it is planed to implement a second trigger stage, which is supposed to make use of the transmitted signal amplitude information. Including for example the correlation between the corresponding amplitudes in the inner and outer ring of the detector into the trigger decision, the purity of the trigger signal is supposed to increase dramatically [11].

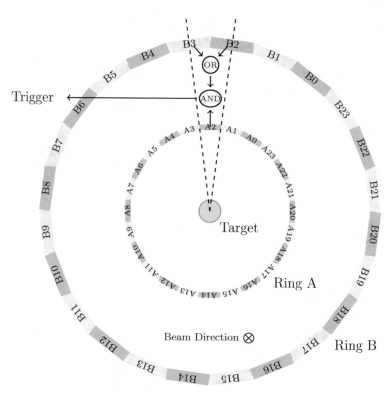

Fig. 3.7 Schematic front view of the CAMERA detector, which illustrates the first stage of the trigger decision (picture adapted from Ref. [9])

3.6.3 The Random Trigger

The random trigger signal is generated by a radioactive $^{22}_{11}$Na source. To avoid a contamination of the signal by beam particle interactions, it is placed in great distance to the beam line. A clean experimental signature of the dominant β^+ decay of $^{22}_{11}$Na to the exited $^{10}_{22}$Ne* state is achieved by a coincidence measurement of two photons. These photons are produced as the decay positron annihilates with a nearby low energetic electron. The logical signal characterising the two photon coincidence is connected to the COMPASS trigger control system via a 1 km long cable. The random trigger is crucial for the measurement of the beam flux, as it will be discussed in Sect. 5.2.

3.7 Data Acquisition and Reconstruction

The data acquisition system realises the readout of over 250,000 detector channels. Until the end of 2012 this was achieved by a modular design, according to Fig. 3.8, which will be outlined in the following.

In order to ensure the best possible signal integrity, the first stage of the readout electronics is placed as close as possible to the respective detector. Within this stage the analogue signals of the individual detector channels are digitised. It comprises mainly TDC[7] and ADC modules.

The second stage of the data acquisition is connected to the trigger control system. To guarantee a synchronous digitisation, it provides the global 38.88 MHz readout clock to the ADC and TDC modules. But it also collects and serialises the digitised information of the individual detector channels within a certain time window with respect to the trigger signal. The first and second stage are either realised on a single readout module or physically separated into two modules.

The serialised data, which is already sorted in an event by event order according to the S-LINK protocol [12], is then either directly transmitted to the spillbuffer computers or multiplexed in an additional stage via the SMUX or TIGER[8] modules. The transmission to the spill buffer computers, which are located more than 50 m away, is achieved by glass fibres with a theoretical data rate of 160 MB/s (in reality <100 MB/s).

The TIGER module was used for the first time in 2012 as a multiplexer. It is capable of concentrating the data of up to 18 GANDALF modules. For the readout of the CAMERA detector it is multiplexing the data of 12 GANDALF modules and thus allows for the readout of 96 channels via a single SLINK fibre. In comparison, the SMUX module is capable to concentrate the data of up to four modules.

The data received by the spillbuffer computers is passed to so called event builder computers. A single event builder receives the information of a complete event of all detector channels of the experiment, which is sorted into one data package and transmitted via Gigabit LAN[9] to the CERN main area for long term storage.

The analysis of the raw data, stored on CASTOR,[10] is performed in several steps. First the raw data is decoded using the DDD[11] library and then processed for track and vertex reconstruction as well as particle identification by the reconstruction software CORAL.[12] This step is commonly called the data production stage. In case of the determination of a charged track this comprises the track reconstruction, using a Kalman filter algorithm [14]. The charged tracks are then combined within a vertex fit, which is also making use of the Kalman filter technology. The Kalman filter has the purpose to decide wheter certain tracks (hits) belong to the same vertex (track). The

---

[7]Time to **D**igital **C**onverter.

[8]**T**rigger **I**mplementation for **G**ANDALF **E**lectronic **R**eadout [13].

[9]**L**ocal **A**rea **N**etwork.

[10]CERN **A**dvanced **STOR**age Manager.

[11]**DAQ D**ata **D**ecoding.

[12]**CO**mpass **R**econstruction and Ana**L**ysis [1].

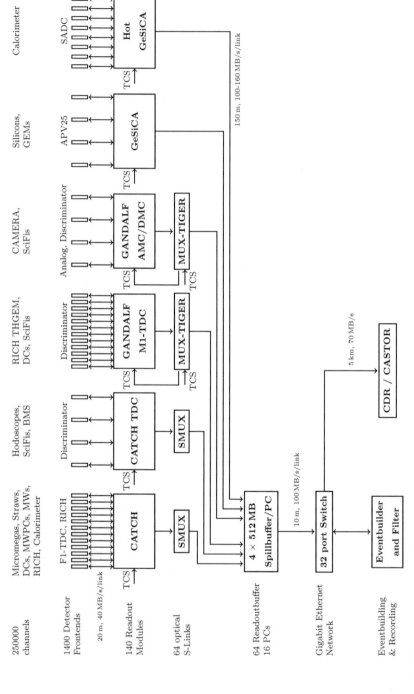

Fig. 3.8 Illustration of the different stages of the data acquisition system used for the 2012 pilot run (picture adapted from Ref. [1])

reconstructed charged tracks and vertices together with the reconstructed calorimeter cluster information and specific information related to particle identification as well as the respective estimated uncertainties on this information are stored in the mDST[13] format. The final data analysis is performed on the mDST level, using the software PHAST.[14] PHAST provides a tool-kit for often used functions to access the information stored inside the mDST files on the basis of the ROOT[15] software framework.

References

1. COMPASS Collaboration, P. Abbon et al., The COMPASS experiment at CERN. Nucl. Instrum. Meth. A **577**, 455–518 (2007). https://doi.org/10.1016/j.nima.2007.03.026
2. E. Bielert et al., A 2.5 m long liquid hydrogen target for COMPASS. Nucl. Instrum. Meth. A **746**, 20–25 (2014). https://doi.org/10.1016/j.nima.2014.01.067
3. T. Szameitat, New Geant4-based Monte Carlo Software for the COMPASS-II Experiment at CERN. Dissertation. Albert Ludwigs Universität Freiburg (2017). https://doi.org/10.6094/UNIFR/11686
4. P. Jörg, Untersuchung von Algorithmen zur Charakterisierung von Photomultiplierpulsen in Echtzeit, Diploma thesis, Albert Ludwigs Universität Freiburg (2013)
5. COMPASS Collaboration, P. Abbon et al., Particle identification with COMPASS RICH-1. Nucl. Instrum. Meth. **631**, 26–39 (2011). https://doi.org/10.1016/j.nima.2010.11.106
6. M. Zimmermann, Untersuchungen mit dem RICH Detektor am COMPASS experiment, Bachelor thesis, Albert Ludwigs Universität Freiburg (2015)
7. A. Ferrero, Artistic picture of the CAMERA detector, (2014), private communications
8. C. Bernet et al., The COMPASS trigger system for muon scattering. Nucl. Instrum. Meth. A **550**, 217–240 (2005). https://doi.org/10.1016/j.nima.2005.05.043
9. T. Baumann, Entwicklung einer Schnittstelle zur Übertragung von Pulsinformationen von einem Rückstoßdetektor an ein digitales Triggersystem, Diploma thesis, Albert Ludwigs Universität Freiburg (2013)
10. M. Gorzellik, Entwicklung eines digitalen Triggersystems für Rückstoßproton Detektoren, Diploma thesis, Albert Ludwigs Universität Freiburg (2013)
11. K. Schaffer, Development of a FPGA based trigger module for the COMPASS-II experiment, Master thesis, Albert Ludwigs Universität Freiburg (2017)
12. H.C. van der Bij et al., S-LINK, a data link interface specification for the LHC era. IEEE T. Nucl. Sci. **44**, 398–402 (1997). https://doi.org/10.1109/23.603679
13. S. Schopferer, An FPGA-based trigger processor for a measurement of deeply virtual compton scattering at the COMPASS-II Experiment, Dissertation, Albert Ludwigs Universität Freiburg (2013), https://freidok.uni-freiburg.de/data/9274, URN: urn:nbn:de:bsz:25-opus-92742
14. R. Fruhwirth, Application of Kalman filtering to track and vertex fitting. Nucl. Instrum. Meth. A **262**, 444–450 (1987). https://doi.org/10.1016/0168-9002(87)90887-4
15. S. Gerassimov, PHysics Analysis Software Tools, http://ges.home.cern.ch/ges/phast
16. R. Brun, Fons Rademakers, ROOT—An object oriented data analysis framework. Nucl. Instrum. Meth. A **389**, 81–86 (1997), proceedings AIHENP'96 Workshop, Lausanne, Sep. 1996, see also: http://root.cern.ch/

---

[13] **m**ini **D**ata **S**ummary **T**rees.

[14] **PH**ysics **A**nalysis **S**oftware **T**ools [15]

[15] ROOT really means the "roots" for end-users applications [16].

Chapter 4
The Kinematically Constrained Fit

The measurements of exclusive single photon or hard exclusive meson production at the COMPASS-II experiment are in general over-constrained. This can be exploited to improve the resolution on the measured kinematic quantities by the usage of a kinematically constrained fit. The chapter is supposed to describe the procedure of a kinematically constrained fit and to give an overview of its application to the experimental situation at COMPASS-II. After a short introduction to the basic mathematical framework, it is shown how the measurements of charged particles, neutral particles and a recoiled target particle are introduced into the procedure and which kind of constraints can be applied.

Later, in Sect. 5.1 the kinematic fitting procedure is applied to the beam and spectrometer measurements of exclusive muoproduction of a ρ^0, to predict the kinematics of the recoiled proton. This allows for the calibration of the longitudinal positions of the scintillators of the CAMERA detector. Finally, in Sect. 7 the full potential of the kinematic fit is exploited. It is applied to the beam, spectrometer and CAMERA measurement of exclusive single photon production.

4.1 Mathematical Description

Technically, a kinematic fit is a constrained minimisation of a scalar function $\chi^2(\vec{k})$ for a set of non-linear constraints of the form $\vec{g}(\vec{k}, \vec{h}) = \vec{0}_I$, while:

$$\vec{g} \in \mathbb{R}^I, \ \vec{k} \in \mathbb{R}^J, \ \vec{h} \in \mathbb{R}^L.$$

The real vector space of dimension M is depicted by $\mathbb{R}^M$ with its neutral element given by $\vec{0}_M$. Using the abbreviation:

$$\Delta\vec{k} = \vec{k} - \vec{k}_{init},$$

© Springer International Publishing AG, part of Springer Nature 2018
P. Jörg, *Exploring the Size of the Proton*, Springer Theses,
https://doi.org/10.1007/978-3-319-90290-6_4

where $\vec{k}_{init}$ denotes a vector of measured values with its corresponding covariance matrix $\hat{C}$, the least squares function can be written as:

$$\chi^2\left(\vec{k}\right) := \Delta\vec{k}^T \hat{C}^{-1} \Delta\vec{k}. \tag{4.1}$$

Its minimisation with respect to the constraints $\vec{g}\left(\vec{k}, \vec{h}\right)$ can be summarised in a minimisation of the Lagrange function:

$$L\left(\vec{k}, \vec{\lambda}\right) = \chi^2\left(\vec{k}\right) + 2\sum_{i=1}^{I} \lambda_i g_i\left(\vec{k}, \vec{h}\right).$$

This is an application of the Lagrange multiplier method, for which the following set of non-linear equations has to be solved:

$$\frac{\partial L\left(\vec{k}, \vec{\lambda}\right)}{\partial \lambda_i} = 0 \text{ (derivatives w.r. to Lagrange multipliers),}$$

$$\frac{\partial L\left(\vec{k}, \vec{\lambda}\right)}{\partial k_j} = 0 \text{ (derivatives w.r. to measured parameters),}$$

$$\frac{\partial L\left(\vec{h}, \vec{\lambda}\right)}{\partial h_l} = 0 \text{ (derivatives w.r. to unmeasured parameters),}$$

$$\forall\, i \in \{1, \ldots, I\};\ \forall\, j \in \{1, \ldots, J\};\ \forall\, l \in \{1, \ldots, L\}.$$

The set of equations can be linearised by a Taylor approximation of the constraints:

$$g_i\left(\vec{k}^{(n+1)}, \vec{h}^{(n+1)}\right) \approx g_i\left(\vec{k}^{(n)}, \vec{h}^{(n)}\right) + \sum_{j=1}^{J} \left.\frac{\partial g_i}{\partial k_j}\right|_{\left(\vec{k}^{(n)}, \vec{h}^{(n)}\right)} \left(\Delta k_j^{(n+1)} - \Delta k_j^{(n)}\right)$$

$$+ \sum_{l=1}^{L} \left.\frac{\partial g_i}{\partial h_l}\right|_{\left(\vec{k}^{(n)}, \vec{h}^{(n)}\right)} \left(\Delta h_l^{(n+1)} - \Delta h_l^{(n)}\right) = 0.$$

The measured and unmeasured quantities of iteration (n), are given by:

$$\Delta k_j^{(n)} = \left(k_j^{(0)} - k_j^{(n)}\right) = \left(k_{init,j} - k_j^{(n)}\right),$$

and respectively:

$$\Delta h_l^{(n)} = \left(h_l^{(0)} - h_l^{(n)}\right) = \left(h_{\approx,l} - h_l^{(n)}\right).$$

The starting points of the iterative procedure are denoted by the measured quantities $\vec{k}^0 := \vec{k}_{init}$ and by estimates of the unmeasured quantities $\vec{h}^0 := \vec{h}_{\approx}$. In more convenient matrix notation the linearisation reads:

$$\vec{g}_i^{(n+1)} = \vec{g}_i^{(n)} + \hat{K}^{(n)}\left(\Delta\vec{k}^{(n+1)} - \Delta\vec{k}^n\right) + \hat{T}^{(n)}\left(\Delta\vec{h}^{(n+1)} - \Delta\vec{h}^{(n)}\right)$$

$$= \hat{K}^{(n)}\Delta\vec{k}^{(n+1)} + \hat{T}^{(n)}\Delta\vec{h}^{(n+1)} - \vec{c}^{(n)} = 0,$$

with the quantities of iteration (n) given by:

$$\hat{K}^{(n)} = \left.\frac{\partial\vec{g}}{\partial\vec{k}}\right|_{(\vec{k}^{(n)},\vec{h}^{(n)})}, \quad \hat{T}^{(n)} = \left.\frac{\partial\vec{g}}{\partial t}\right|_{(\vec{k}^{(n)},\vec{h}^{(n)})}, \quad \vec{c}^{(n)} = \hat{K}^{(n)}\Delta\vec{k}^{(n)} + \hat{T}^{(n)}\Delta\vec{h}^{(n)} - \vec{g}^{(n)},$$

(4.2)

while matrices are denoted with a hat.

Now L can be written as:

$$L = \left(\Delta\vec{k}^{(n+1)}\right)^T \hat{C}^{-1}\Delta\vec{k}^{(n+1)} + 2\vec{\lambda}^T\left(\hat{K}^{(n)}\Delta\vec{k}^{(n+1)} + \hat{T}^{(n)}\Delta\vec{h}^{(n+1)} - \vec{c}^{(n)}\right),$$

and the equations to solve are linear in terms of $\vec{\lambda}$, $\Delta\vec{k}^{(n+1)}$ and $\Delta\vec{h}^{(n+1)}$:

$$\hat{C}^{-1}\Delta\vec{k}^{(n+1)} + \left(\hat{K}^{(n)}\right)^T\vec{\lambda} = 0, \quad (J \text{ equations for the measured parameters}),$$

$$\left(\hat{T}^{(n)}\right)^T\vec{\lambda} = 0, \qquad\qquad (L \text{ equations for the unmeasured parameters}),$$

$$\hat{K}^{(n)}\Delta\vec{k}^{(n+1)} + \hat{T}^{(n)}\Delta\vec{h}^{(n+1)} - \vec{c}^{(n)} = 0, \qquad (I \text{ equations for the constraints}).$$

Solving this linearised system of equations yields:

$$\Delta\vec{k}^{(n+1)} = \hat{C}\left(\hat{K}^{(n)}\right)^T \hat{C}_K^{(n)}\left[1 - \hat{T}^{(n)}\left(\hat{C}_T^{(n)}\right)^{-1}\hat{C}_K^{(n)}\left(\hat{T}^{(n)}\right)^T \hat{C}_K^{(n)}\right]\vec{c}^{(n)},$$

$$\Delta\vec{h}^{(n+1)} = \left(\hat{C}_T^{(n)}\right)^{-1}\left(\hat{T}^{(n)}\right)^T \hat{C}_K^{(n)}\vec{c}^{(n)},$$

$$\vec{\lambda}^{(n+1)} = \hat{C}_K^{(n)}\left[\hat{T}^{(n)}\left(\hat{C}_T^{(n)}\right)^{-1}(\hat{T}^{(n)})^T \hat{C}_K^{(n)} - 1\right]\vec{c}^{(n)},$$

while $\hat{C}_K^{(n)} = \left[\hat{K}^{(n)}\hat{C}\left(\hat{K}^{(n)}\right)^T\right]^{-1}$ and $\hat{C}_T^{(n)} = \left(\hat{T}^{(n)}\right)^T \hat{C}_K^{(n)}\hat{T}^{(n)}$ have been used for better readability.

The full covariance matrix for the vector of minimised parameters $\left(\vec{k}^{(n+1)}, \vec{h}^{(n+1)}\right)$ is derived by Gaussian error propagation to be:

$$\hat{C}_f^{(n)} = \begin{pmatrix} \left(\hat{C}_{11}^{(n)}\right) & \left(\hat{C}_{21}^{(n)}\right)^T \\ \left(\hat{C}_{21}^{(n)}\right) & \left(\hat{C}_{22}^{(n)}\right) \end{pmatrix},$$

with the abbreviations:

$$\hat{C}_{11}^{(n)} = \hat{C}\left[1 - \left(\hat{K}^{(n)}\right)^T \hat{C}_K^{(n)} \hat{K}^{(n)} \hat{C} + \left(\hat{K}^{(n)}\right)^T \hat{C}_K^{(n)} \hat{T}^{(n)} \left(\hat{C}_T^{(n)}\right)^{-1} \left(\hat{T}^{(n)}\right)^T \hat{C}_K^{(n)} \hat{K}^{(n)} \hat{C}\right],$$

$$\hat{C}_{21}^{(n)} = -\left(\hat{C}_T^{(n)}\right)^{-1} \left(\hat{T}^{(n)}\right)^T \hat{C}_K^{(n)} \hat{K}^{(n)} \hat{C},$$

and

$$\hat{C}_{22}^{(n)} = \left(\hat{C}_T^{(n)}\right)^{-1}.$$

The convergence of the procedure is achieved in case the following two criteria are satisfied:

$$\frac{\chi^2\left(\vec{k}^{(n+1)}\right) - \chi^2\left(\vec{k}^{(n)}\right)}{ndf} < \epsilon_\chi,$$

$$\sum_{i=1}^{I} \left| g_i\left(\vec{k}^{(n+1)}, \vec{h}^{(n+1)}\right) \right| < \epsilon_g,$$

The abbreviation ndf denotes the number of degrees of freedom, which is given by the difference between the number of constraints I and the number of free parameters J. The quantities ϵ_χ and ϵ_g denote two real parameters.

A comprehensive and more detailed description of the mathematical framework is given in [1]. The procedure developed during this thesis is making use of the publicly available software of [2], which provides the minimisation procedure described in this section and a basic set of momentum, energy and mass constraints.

4.2   Definition of the Input Covariance Matrix

The input covariance $\hat{C}$ in Eq. 4.1 is a block diagonal matrix:

$$\hat{C} = \begin{pmatrix} \hat{C}_1 & 0 & \dots & 0 \\ 0 & \hat{C}_2 & \dots & 0 \\ \cdot & \cdot & \cdot & \cdot \\ \cdot & \cdot & \cdot & \cdot \\ \cdot & \cdot & \cdot & \cdot \\ 0 & 0 & \dots & \hat{C}_N \end{pmatrix}. \tag{4.3}$$

For each of the N particles, which enter the kinematic fit, a matrix $\hat{C}_i$ enters the diagonal of $\hat{C}$. Hence, the measured track parameters of one particle are assumed to be independent with respect to the ones of another particle. It shall be emphasised that a correlation amongst the track parameters of the different particles will appear in case the parameters are already the result of a vertex fit. Thus, in the following the raw track parameters, not corrected by a vertex fit, are used and the vertex fit is incorporated in the kinematic fit by adding vertex constraints.

Correlations amongst the determined track parameters of one particle are taken into account within the corresponding covariance matrices $\hat{C}_i$. Due to the experimental situation the dimension and the most appropriate choice of coordinates for the $\hat{C}_i$ can differ. The three cases described in the following three sections should be distinguished.

4.3 Treatment of Charged Tracks

The helix of a charged track is represented in the reconstruction software CORAL as:

$$\vec{S} = \begin{pmatrix} x \\ y \\ X \\ Y \\ |\vec{p}|^{-1} \end{pmatrix}.$$

The quantities X and Y are short-handed for $\frac{dx}{dz}$ and $\frac{dy}{dz}$, while x and y are the transverse coordinates of the charged particle with momentum $\vec{p}$ at a given longitudinal coordinate z. A covariance matrix $\hat{C}$ for the coordinates of this track representation is derived by a track fit during the reconstruction process and available in the analysis software PHAST. In order to formulate momentum conservation constraints in Cartesian coordinates, a transformation into the track representation,

$$\vec{S}' = \begin{pmatrix} x \\ y \\ p_x \\ p_y \\ p_z \end{pmatrix}, \tag{4.4}$$

has to be performed. The relation between the two representations is the following:

$$\vec{S}' = \begin{pmatrix} x \\ y \\ 0 \\ 0 \\ 0 \end{pmatrix} + \frac{|\vec{p}|}{\sqrt{1 + X^2 + Y^2}} \begin{pmatrix} 0 \\ 0 \\ X \\ Y \\ 1 \end{pmatrix}. \tag{4.5}$$

Hence the Jacobi matrix $\hat{J} = \frac{\partial \vec{S'}}{\partial \vec{S}}$, describing the transformation between the two representations, is given by:

$$\hat{J} = \frac{1}{w^3|\vec{p}|^{-1}} \begin{pmatrix} w^3|\vec{p}|^{-1} & 0 & 0 & 0 & 0 \\ 0 & w^3|\vec{p}|^{-1} & 0 & 0 & 0 \\ 0 & 0 & w^2 - X^2 & -XY & -w^3 p_x \\ 0 & 0 & -XY & w^2 - Y^2 & -w^3 p_y \\ 0 & 0 & -w|\vec{p}|^{-1} p_x & -w|\vec{p}|^{-1} p_y & -w^3 p_z \end{pmatrix},$$

while the abbreviation $w = \sqrt{1 + X^2 + Y^2}$ has been used. The covariance matrix $\hat{C}'$ in the Cartesian representation is finally related to the covariance matrix $\hat{C}$, available in the analysis software, by basis transformation:

$$\hat{C}' = \hat{J}\hat{C}\hat{J}^T. \tag{4.6}$$

The initial quantity $\vec{S}$ is given by an extrapolation of the track parameters at the z-position of the first measured point to the z-position of the interaction vertex. The z-position of the interaction vertex is taken from the vertex fit of the reconstruction software CORAL. The extrapolation through the magnetic-field is performed by the analysis software PHAST. It takes into account energy loss effects and multiple scattering uncertainties inside the covariance matrix $\hat{C}$. It shall be emphasised that the track parameters at the first measured point are not corrected by the vertex fit, performed inside CORAL. The z-position of the vertex given by CORAL is simply used in order to have a good estimate of energy loss effects and multiple scattering uncertainties within $\vec{S}$ and $\hat{C}$.

The advantage of this approach is that $\vec{S}'$ and $\hat{C}'$ from Eqs. 4.5 and 4.6 can now be used to find a common vertex within the kinematic fit procedure, using a straight line approximation for the vertex constraints inside the field free region. This allows to calculate the derivatives of Eq. 4.2 with respect to the vertex constraints analytically.

To summarise, the input parameters $\vec{S}^{in}$ for each charged particle into the kinematic fit procedure are given by $\vec{S}'$:

$$\vec{S}^{in} := \vec{S}'. \tag{4.7}$$

The input covariance matrix $\hat{C}^{meas}$ is given by:

$$\hat{C}^{meas} := \hat{C}', \tag{4.8}$$

It is $\hat{C}^{meas}$ which enters the diagonal in Eq. 4.3.

4.4 Treatment of Photons

In case of a neutral cluster[1] the cluster position $\vec{r}$ is reconstructed inside one of the three electromagnetic calorimeters. This is performed by the clusterisation algorithms inside the reconstruction software CORAL. Furthermore, as a result of a final calibration of the calorimeters, the photon momentum $|\vec{p}|$ at the interaction vertex is also available. Due to the experimental situation it is then appropriate to choose a track parametrisation as follows:

$$\vec{S} = \begin{pmatrix} x \\ y \\ |\vec{p}| \\ \theta_p \\ \phi_p \end{pmatrix}.$$

The quantities x and y denote the transverse coordinates of the cluster at the z-position of the respective calorimeter and ϕ_p, θ_p the azimuthal and polar angle of the cluster momentum $\vec{p}$. Since the constraints are formulated in Cartesian coordinates, one has to perform the basis transformation into the Cartesian representation $\vec{S}'$. The relation between the two representations is the following:

$$\vec{S}' = \begin{pmatrix} x \\ y \\ 0 \\ 0 \\ 0 \end{pmatrix} + \begin{pmatrix} 0 \\ 0 \\ |\vec{p}| \sin \theta_p \cos \phi_p \\ |\vec{p}| \sin \theta_p \sin \phi_p \\ |\vec{p}| \cos \theta_p \end{pmatrix}.$$

The Jacobi matrix $\hat{J} = \frac{\partial \vec{S}'}{\partial \vec{S}}$, describing the transformation between the two representations, is thus given by:

$$\hat{J} = \begin{pmatrix} 1 & 0 & 0 & 0 & 0 \\ 0 & 1 & 0 & 0 & 0 \\ 0 & 0 & \sin \theta_p \cos \phi_p & |\vec{p}| \cos \theta_p \cos \phi_p & -|\vec{p}| \sin \theta_p \sin \phi_p \\ 0 & 0 & \sin \theta_p \sin \phi_p & |\vec{p}| \cos \theta_p \sin \phi_p & |\vec{p}| \sin \theta_p \cos \phi_p \\ 0 & 0 & \cos \theta_p & -|\vec{p}| \sin \theta_p & 0 \end{pmatrix}.$$

At this point one could imagine to proceed as described in Sect. 4.3. However, the situation is different since in case of a neutral cluster the photon momentum is not

---

[1]A neutral cluster is defined as a reconstructed calorimeter cluster with no charged track pointing to its location in the electromagnetic calorimeter.

fully measured and the unmeasured quantities θ_p and ϕ_p have to be determined by the kinematic fit. Thus, the derivatives in Eq. 4.2 for a certain constraint i are evaluated using the chain rule:

$$\frac{\partial g_i}{\partial \vec{S}} = \frac{\partial g_i}{\partial \vec{S'}} \frac{\partial S'}{\partial \vec{S}} = \frac{\partial g_i}{\partial \vec{S'}} \hat{J} =: \vec{b},$$

and there are only derivatives with respect to the Cartesian representation S left to calculate:

$$\frac{\partial g_i}{\partial \vec{S'}} = \left(\frac{\partial g_i}{\partial x}, \frac{\partial g_i}{\partial y}, \frac{\partial g_i}{\partial p_x}, \frac{\partial g_i}{\partial p_y}, \frac{\partial g_i}{\partial p_z} \right).$$

The derivatives in Eq. 4.2 are now explicitly given in case of the measured quantities $\vec{k} = (x, y, |\vec{p}|)^T$ by:

$$\frac{\partial g_i}{\partial \vec{k}} = \left(\frac{\partial g_i}{\partial x}, \frac{\partial g_i}{\partial y}, \frac{\partial g_i}{\partial |\vec{p}|} \right) = \left(b_1, b_1, b_3 \right),$$

and in case of the unmeasured quantities $\vec{h} = (\phi, \theta)^T$ by:

$$\frac{\partial g_i}{\partial \vec{h}} = \left(\frac{\partial g_i}{\partial \theta_p}, \frac{\partial g_i}{\partial \phi_p} \right) = \left(b_4, b_5 \right).$$

The input parameters to the kinematic fitting procedure $\vec{S}^{meas}$ for each neutral cluster are then given by the measured quantities according to:

$$\vec{S}^{meas} := \begin{pmatrix} x \\ y \\ |\vec{p}| \end{pmatrix}, \tag{4.9}$$

while the input covariance matrix $\hat{C}^{meas}$, entering the diagonal in Eq. 4.3, has the following form:

$$\hat{C}^{meas} := \begin{pmatrix} \hat{C}_{xy} & 0 \\ 0 & \sigma_{|\vec{p}|}^2 \end{pmatrix}. \tag{4.10}$$

The quantity $\hat{C}_{xy}$ takes the correlation between the x and y position of the cluster into account and is given as a result of the clusterisation algorithms for the different calorimeters. The photon momentum resolution is denoted by $\sigma_{|\vec{p}|}$. It is also given by the clusterisation algorithms. The correlation between $|\vec{p}|$ and x or y is assumed to vanish.

The parameters ϕ_p and θ_p are free parameters to be determined by the kinematic fit and are closely connected with the fact that the photon will be constrained to have its origin at the interaction vertex.

4.5 Treatment of the Recoiled Target Particle

For a recoiled target particle, measured by the CAMERA detector, the hit positions $\vec{r}_A = (r_A, \phi_A, z_A)$ and $\vec{r}_B = (r_B, \phi_B, z_B)$ inside the inner and outer ring of scintillators are known, while the momentum $|\vec{p}|$ is reconstructed by a time of flight and distance of flight measurement. An appropriate description of the measurement, reflecting the barrel shaped detector, is thus given by:

$$
\vec{S} = \begin{pmatrix} r_A \\ \phi_A \\ z_A \\ r_B \\ \phi_B \\ z_B \\ |\vec{p}| \\ \theta_p \\ \phi_p \end{pmatrix} . \tag{4.11}
$$

The quantities ϕ_p and θ_p denote the azimuthal and polar angle of the proton momentum $\vec{p}$. The relation with a Cartesian representation $\vec{S}'$ is the following:

$$
\vec{S}' = \begin{pmatrix} r_A \cos \phi_A \\ r_A \sin \phi_A \\ z_A \\ \vec{0}_6 \end{pmatrix} + \begin{pmatrix} \vec{0}_3 \\ r_B \cos \phi_B \\ r_B \sin \phi_B \\ z_B \\ \vec{0}_3 \end{pmatrix} + \begin{pmatrix} \vec{0}_6 \\ |\vec{p}| \sin \theta_p \cos \phi_p \\ |\vec{p}| \sin \theta_p \sin \phi_p \\ |\vec{p}| \cos \theta_p \end{pmatrix} .
$$

The Jacobi matrix $\hat{J} = \frac{\partial \vec{S}'}{\partial \vec{S}}$, describing the transformation between the two representations, is given by:

$$
\hat{J} = \begin{pmatrix} \hat{J}_A & 0 & 0 \\ 0 & \hat{J}_B & 0 \\ 0 & 0 & \hat{J}_p \end{pmatrix},
$$

while the following abbreviations have been used:

$$
\hat{J}_{A;B} = \begin{pmatrix} \cos(\phi_{A;B}) & -r_{A;B} \sin(\phi_{A;B}) & 0 \\ \sin(\phi_{A;B}) & r_{A;B} \cos(\phi_{A;B}) & 0 \\ 0 & 0 & 1 \end{pmatrix},
$$

$$
\hat{J}_p = \begin{pmatrix} \sin \theta_p \cos \phi_p & |\vec{p}| \cos \theta_p \cos \phi_p & -|\vec{p}| \sin \theta_p \sin \phi_p \\ \sin \theta_p \sin \phi_p & |\vec{p}| \cos \theta_p \sin \phi_p & |\vec{p}| \sin \theta_p \cos \phi_p \\ \cos \theta_p & -|\vec{p}| \sin \theta_p & 0 \end{pmatrix} .
$$

As described in Sect. 4.4 the derivatives in Eq. 4.2 for a certain constraint i are evaluated using the chain rule:

$$\frac{\partial g_i}{\partial \vec{S}} = \frac{\partial g_i}{\partial \vec{S'}} \frac{\partial S'}{\partial \vec{S}} = \frac{\partial g_i}{\partial \vec{S'}} \hat{J} =: \vec{b},$$

and there are only derivatives with respect to the Cartesian representation S left to calculate:

$$\frac{\partial g_i}{\partial \vec{S'}} = \left(\frac{\partial g_i}{\partial x_A}, \frac{\partial g_i}{\partial y_A}, \frac{\partial g_i}{\partial z_A}, \frac{\partial g_i}{\partial x_B}, \frac{\partial g_i}{\partial y_B}, \frac{\partial g_i}{\partial z_B}, \frac{\partial g_i}{\partial p_x}, \frac{\partial g_i}{\partial p_y}, \frac{\partial g_i}{\partial p_z} \right).$$

The derivatives are then explicitly given in case of the measured quantities $\vec{k} = (r_A, \phi_A, z_A, r_B, \phi_B, z_B, |\vec{p}|)^T$ by:

$$\frac{\partial g_i}{\partial \vec{k}} = \left(\frac{\partial g_i}{\partial r_A}, \frac{\partial g_i}{\partial \phi_A}, \frac{\partial g_i}{\partial z_A}, \frac{\partial g_i}{\partial r_B}, \frac{\partial g_i}{\partial \phi_B}, \frac{\partial g_i}{\partial z_B}, \frac{\partial g_i}{\partial |\vec{p}|} \right) = \left(b_1, \ldots, b_7 \right),$$

and in case of the unmeasured quantities $\vec{h} = (\theta_p, \phi_p)^T$ by:

$$\frac{\partial g_i}{\partial \vec{h}} = \left(\frac{\partial g_i}{\partial \phi_p}, \frac{\partial g_i}{\partial \theta_p} \right) = \left(b_8, b_9 \right).$$

The parameters ϕ_p and θ_p of Eq. 4.11 are free parameters to be determined by the kinematic fit and not part of the input parameters.

The input parameters S^{meas} to the kinematic fitting procedure for each track, detected inside the CAMERA detector, are then given according to:

$$S^{meas} := \begin{pmatrix} r_A \\ \phi_A \\ z_A \\ r_B \\ \phi_B \\ z_B \\ |\vec{p}| \end{pmatrix}. \tag{4.12}$$

The input covariance matrix $\hat{C}^{meas}$, entering the diagonal in Eq. 4.3, has the following form:

$$C^{meas} := \begin{pmatrix} \hat{C}_A & 0 & 0 \\ 0 & \hat{C}_B & 0 \\ 0 & 0 & \sigma_p^2 \end{pmatrix}. \tag{4.13}$$

The quantity $\hat{C}_{A,B}$ is given by:

$$\hat{C}_{A,B} = \begin{pmatrix} \sigma_r^2(A; B) & 0 & 0 \\ 0 & \sigma_\phi^2(A; B) & 0 \\ 0 & 0 & \sigma_z^2(A; B) \end{pmatrix}.$$

The values $\sigma_r(A; B)$ denote the uncertainties on the shortest distance of a ring A or B counter to the centre of the CAMERA detector. They have been chosen to be at the order of the width of the two different types of counters:

$$\sigma_r(A) = 0.4\,\text{cm}; \ \sigma_r(B) = 5\,\text{cm}.$$

The values $\sigma_\phi(A; B)$ denote the azimuthal uncertainty on the measurement, due to the discrete number of counters of ring A and B. They have been chosen to be:

$$\sigma_\phi(A; B) = \frac{2\pi}{24\sqrt{12}}\,\text{rad}.$$

The uncertainties $\sigma_z(A; B)$ denote the resolutions of the longitudinal hit positions of ring A and B. They are given by:

$$\sigma_z(A) = 4.1\,\text{cm}; \ \sigma_z(B) = 2.9\,\text{cm}. \tag{4.14}$$

The quantity $\sigma_z(A)$ is the same as it is stated in Eq. 5.7, which corresponds to the assumption that the resolution on the interaction vertex and the resolution of ring B have a negligible impact compared to the "true" resolution of ring A.

The quantity $\sigma_z(B)$ is changed slightly with respect to Eq. 5.6, in order to account for the fact that Eq. 5.6 states the resolution of ring B with respect to the spectrometer and beam measurement. The estimate for the ring B resolution is chosen such that a consistent picture between data and Monte Carlo with respect to Figs. 6.9 and 6.15 arises.

Finally, the value of σ_p in Eq. 4.13 denotes the experimental resolution on the magnitude of the momentum of the recoiled proton. It is estimated by means of a Monte Carlo simulation. A detailed description of the introduction of the CAMERA detector to the Monte Carlo simulations is given in Sect. 6.1. In this context the introduction procedure essentially corresponds to a transformation of $\sigma_z(A; B)$ to a resolution of the corresponding time stamps at the up- and downstream side of each counter. Since the time of flight of a recoiled particle is calculated directly from these time-stamps, one gains insight into the momentum resolution within the simulations.

Figure 4.1 shows the resolution on the magnitude of the proton momentum using a single photon Monte Carlo yield. From this simulation σ_p is estimated according to the red line of Fig. 4.1.

Fig. 4.1 Measured momentum resolution of protons. The quantity $\vec{p}$ denotes the reconstructed proton momentum, given by the hit information of the CAMERA detector, the quantity t the square of the four-momentum transfer to the proton. These data have been taken from Monte Carlo simulations of single photon production

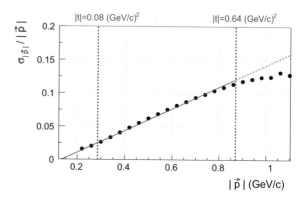

The fact that the red line overestimates the momentum resolution of CAMERA in the region of large proton momenta or large values of $|t|$, the magnitude of the four momentum transfer to the proton squared, is not too much of a problem. As it will be shown in Sect. 6.3, the determination of $|t|$ above $0.4\,(\text{GeV/c})^2$ by a kinematic fit which combines beam spectrometer and CAMERA measurements is completely dominated by the beam and spectrometer measurement. Thus, a slight overestimation of the resolution of the proton momentum measured by CAMERA corresponds to a negligible loss of precision in the region of large $|t|$. Or to put it in other words, in the region of $|t|$ above $0.4\,(\text{GeV/c})^2$ the resolution on $|t|$ reconstructed by CAMERA is much worse compared to the resolution of the beam and spectrometer measurement. Hence, a slight overestimation of the CAMERA resolution in the region of large $|t|$ will have no noticable impact on a combined beam spectrometer and CAMERA determination.

4.6   Constraints

In this section the different types of constraints are discussed. The constraints are formulated in Cartesian coordinates, while the input quantities and the corresponding uncertainties are given in the experimentally most applicable coordinates. The translation between the different coordinates is explicitly shown in Sects. 4.3–4.5.

4.6.1   Energy and Momentum Constraints

The momentum constraints for N incoming and J outgoing particles have the following form:

$$g_k(\vec{p}^{\,[1]}, \ldots, \vec{p}^{\,[N]}, \vec{p}'^{\,[1]}, \ldots, \vec{p}'^{\,[J]}) = \sum_{n=1}^{n=N}(\vec{p}^{\,[n]})\vec{e}_k - \sum_{j=1}^{j=J}(\vec{p}'^{\,[j]})\vec{e}_k = 0,$$

$$k \in \{1, 2, 3\},$$

where $\vec{e}_k$ denotes the unity vector into the x-, y- or z-direction and the outgoing particles have been denoted with a slash. The derivatives necessary to evaluate Eq. 4.2 are thus given by:

$$\frac{\partial g_k}{\partial p_l^{[n]}} = \delta_{l,k}\, p_l^{[n]} \quad \text{and} \quad \frac{\partial g_k}{\partial p_l'^{[j]}} = -\delta_{l,k}\, p_l'^{[j]}.$$

Denoting the mass of the nth particle by $m^{[n]}$, the energy constraint for N incoming and J outgoing particles has the following form:

$$g_E(E^{[1]}, \ldots, E^{[N]}, E'^{[1]}, \ldots, E'^{[J]}) = \sum_{n=1}^{n=N} E^{[n]} - \sum_{j=1}^{j=J} E'^{[j]} = 0.$$

The derivatives needed in Eq. 4.2 are thus given by:

$$\frac{\partial g_E}{\partial p_l^{[n]}} = \frac{\partial g_E}{\partial E^{[n]}}\frac{\partial E^{[n]}}{\partial p_l^{[n]}} = \frac{\partial E^{[n]}}{\partial p_l^{[n]}} = \frac{p_l^{[n]}c^2}{\sqrt{\sum_{i=1}^3 (p_i^{[n]}c)^2 + (m^{[n]}c^2)^2}} = \frac{p_l^{[n]}c^2}{E^{[n]}},$$

and respectively for an outgoing particle:

$$\frac{\partial g_E}{\partial p_l'^{[j]}} = -\frac{p_l'^{[j]}c^2}{E'^{[j]}}.$$

All remaining derivatives of the above constraints with respect to parameters which are not part of the respective constraint are equal to zero.

4.6.2 Vertex Constraints

The track of a particle in the absence of a magnetic field can be parametrised by a straight line:

$$\vec{r}(\eta) = \vec{a} + \eta\vec{p},$$

while $\vec{a}$ denotes a known point on the track, $\vec{p}$ its momentum and η a free parameter. Writing the equation for the z-component yields:

$$r_z = a_z + \eta p_z \Rightarrow \eta p_z = r_z - a_z. \tag{4.15}$$

For the x-component one finds:

$$r_x = a_x + \eta p_x \Rightarrow r_x p_z = a_x p_z + \eta p_x p_z. \tag{4.16}$$

Inserting ηp_z from Eqs. 4.15 into 4.16, it follows:

$$p_z(r_x - a_x) - p_x(r_z - a_z) = 0. \tag{4.17}$$

An analogue procedure for the y-component yields:

$$p_z(r_y - a_y) - p_y(r_z - a_z) = 0. \tag{4.18}$$

Hence, a point $\vec{r}$ can be found on the track in case Eqs. 4.17 and 4.18 are simultaneously satisfied, or in other words a line is constrained by two planes. In case one imagines to only have one fully measured track, the problem is underconstrained since one can not determine the three components of the vertex with only two equations. As soon as a second track is measured the problem immediately becomes over-constrained and this fact is then commonly used to improve resolutions on the track parameters itself. Thus, for N particles constrained to a common vertex $\vec{v}$ the equations:

$$g_1^{[i]} = p_z^{[i]}(v_x - a_x^{[i]}) - p_x^{[i]}(v_z - a_z^{[i]}) = 0, \tag{4.19}$$

and

$$g_2^{[i]} = p_z^{[i]}(v_y - a_y^{[i]}) - p_y^{[i]}(v_z - a_z^{[i]}) = 0, \tag{4.20}$$

have to be satisfied for all $i \in \{1, \ldots, N\}$.
The derivatives needed in Eq. 4.2 are thus explicitly given by:

$$\frac{\partial g_1^{[i]}}{\partial p_x^{[j]}} = -\delta_{i,j}\,(v_z - a_z^{[i]}), \quad \frac{\partial g_1^{[i]}}{\partial p_y^{[j]}} = 0, \quad \frac{\partial g_1^{[i]}}{\partial p_z^{[j]}} = \delta_{i,j}\,(v_x - a_x^{[i]}),$$

$$\frac{\partial g_1^{[i]}}{\partial a_x^{[j]}} = -\delta_{i,j}\,p_z^{[i]}, \quad \frac{\partial g_1^{[i]}}{\partial a_y^{[j]}} = 0, \quad \frac{\partial g_1^{[i]}}{\partial a_z^{[j]}} = \delta_{i,j}\,p_x^{[i]},$$

$$\frac{\partial g_1^{[i]}}{\partial v_x} = p_z^{[i]}, \quad \frac{\partial g_1^{[i]}}{\partial v_y} = 0, \quad \frac{\partial g_1^{[i]}}{\partial v_z} = -p_x^{[i]},$$

and

$$\frac{\partial g_2^{[i]}}{\partial p_x^{[j]}} = 0, \quad \frac{\partial g_2^{[i]}}{\partial p_y^{[j]}} = -\delta_{i,j}\,(v_z - a_z^{[i]}), \quad \frac{\partial g_2^{[i]}}{\partial p_z^{[j]}} = \delta_{i,j}\,(v_y - a_y^{[i]}),$$

$$\frac{\partial g_2^{[i]}}{\partial a_x^{[j]}} = 0, \quad \frac{\partial g_2^{[i]}}{\partial a_y^{[j]}} = -\delta_{i,j}\,p_z^{[i]}, \quad \frac{\partial g_2^{[i]}}{\partial a_z^{[j]}} = \delta_{i,j}\,p_y^{[i]},$$

$$\frac{\partial g_2^{[i]}}{\partial v_x} = 0, \quad \frac{\partial g_2^{[i]}}{\partial v_y} = p_z^{[i]}, \quad \frac{\partial g_2^{[i]}}{\partial v_z} = -p_y^{[i]},$$

$$\forall\, i, j \in \{1, \ldots, N\}.$$

Similar to Sect. 4.6.1 all remaining derivatives of the respective constraints with respect to parameters which are not part of the constraints, are equal to zero.

4.6.3 Extrapolation Constraints

The extrapolation constraint is very much similar to a vertex constraint. The idea is to assume that a particle originates from the interaction vertex and to constrain it to a set of measured positions on its track. Applying subsequently the following substitutions to the set of Eqs. 4.19 and 4.20:

$$\vec{v} \to \vec{r}^{[i]},$$

$$\vec{a}^{[i]} \to \vec{v},$$

$$\vec{p}^{[i]} \to \vec{p},$$

the extrapolation constraints read:

$$g_1^{[i]} = p_z(r_x^{[i]} - v_x) - p_x(r_z^{[i]} - v_z) = 0,$$

and

$$g_2^{[i]} = p_z(r_y^{[i]} - v_y) - p_y(r_z^{[i]} - v_z) = 0.$$

In this context $\vec{p}$ denotes the momentum of the particle, $\vec{v}$ the interaction vertex, $\vec{r}^{[i]}$ the available reconstructed hit positions with $i \in \{1, \ldots, N\}$ and N the number of available position measurements along the track. For completeness the required derivatives in Eq. 4.2 are listed in case of the interpolation constraints in the following:

$$\frac{\partial g_1^{[i]}}{\partial p_x} = -(r_z^{[i]} - v_z), \quad \frac{\partial g_1^{[i]}}{\partial p_y} = 0, \quad \frac{\partial g_1^{[i]}}{\partial p_z} = (r_x^{[i]} - v_x),$$

$$\frac{\partial g_1^{[i]}}{\partial r_x^{[j]}} = \delta_{i,j}\, p_z, \quad \frac{\partial g_1^{[i]}}{\partial r_y^{[j]}} = 0, \quad \frac{\partial g_1^{[i]}}{\partial r_z^{[j]}} = -\delta_{i,j}\, p_x,$$

$$\frac{\partial g_1^{[i]}}{\partial v_x} = -p_z, \quad \frac{\partial g_1^{[i]}}{\partial v_y} = 0, \quad \frac{\partial g_1^{[i]}}{\partial v_z} = p_x^{[i]},$$

and

$$\frac{\partial g_2^{[i]}}{\partial p_x} = 0, \quad \frac{\partial g_2^{[i]}}{\partial p_y} = -(r_z^{[i]} - v_z), \quad \frac{\partial g_2^{[i]}}{\partial p_z} = (r_y^{[i]} - v_y),$$

$$\frac{\partial g_2^{[i]}}{\partial r_x^{[j]}} = 0, \quad \frac{\partial g_2^{[i]}}{\partial r_y^{[j]}} = \delta_{i,j}\, p_z, \quad \frac{\partial g_2^{[i]}}{\partial r_z^{[j]}} = -\delta_{i,j}\, p_y,$$

$$\frac{\partial g_2^{[i]}}{\partial v_x} = 0, \quad \frac{\partial g_2^{[i]}}{\partial v_y} = -p_z, \quad \frac{\partial g_2^{[i]}}{\partial v_z} = p_y,$$

$$\forall\, i, j \in \{1, \ldots, N\}.$$

Similar to Sects. 4.6.1 and 4.6.2 all remaining derivatives of the respective constraints with respect to parameters which are not part of the constraints, are equal to zero.

The extrapolation constraint is explicitly designed for the experimental situation of the CAMERA detector. Hence, in the following the set of hit positions is given by the reconstructed hit positions inside ring A and B of the CAMERA detector implying $N = 2$.

References

1. R .K. Bock et al., Formulae and methods in experimental data evaluation, with emphasis on high energy physics: articles on statistical and numerical methods. CERN Geneva **3** (1983)
2. T. Goepfert, KinFitter A Kinematic Fit with Constraints. https://github.com/goepfert/KinFitter/wiki/KinFitter---A-Kinematic-Fit-with-Constraints

Chapter 5
The 2012 DVCS Data

The 2012 DVCS run was performed as a pilot run for the dedicated DVCS beam time through 2016 and 2017. The charge and polarisation of the muon beam were changed nine times during five weeks, such that five periods for each beam charge and polarisation are recorded. The data taken with the μ^+ and respectively the μ^- beam will be denoted as the μ^+ or μ^- data yield in the following. Between μ^+ and μ^- data taking periods the magnetic fields of the two spectrometer dipole magnets were inverted. The first part of this chapter is supposed to explain the calibration procedure of the CAMERA detector. The second part deals with the extraction of the luminosity, the application of stability criteria to the data and the determination of the efficiency of the CAMERA detector.

5.1 Calibration of the CAMERA Detector

The CAMERA detector is supposed to measure the momentum and direction of the recoiled target proton, as it is described in Sect. 3.5. During the 2012 data taking it was used for the first time as a part of the COMPASS apparatus. The calibration procedure of the time and distance of flight measurement of the recoiled target particles is described throughout this section.

5.1.1 The Exclusive ρ^0 Sample

In order to calibrate the time of flight and distance of flight measurement of the CAMERA detector, exclusive ρ^0 muoproduction is used. After its production the ρ^0 decays almost instantly into two charged pions. The cross section of the reaction

© Springer International Publishing AG, part of Springer Nature 2018
P. Jörg, *Exploring the Size of the Proton*, Springer Theses,
https://doi.org/10.1007/978-3-319-90290-6_5

$\mu p \rightarrow \mu' p' \rho^0 \rightarrow \mu' p' \pi^+ \pi^-$ is large enough to provide reasonable statistics for a separate calibration of each of the 48 scintillating counters of the CAMERA detector. Due to the exclusive character of the reaction, the measurement of the beam and scattered muon together with the two charged pions allows for a prediction of the momentum and the interaction points of the recoiled proton within the CAMERA detector. The selection of exclusive ρ^0 events is kept close to Ref. [1], while certain criteria of the selection were changed in order to gain statistics (Table 5.1).

Table 5.1 Overview of the selection of exclusive ρ^0 events necessary for the calibration of CAMERA

• Best primary vertex[a] • One incident muon: μ • One outgoing charged track with same charge than incoming: μ' • Two outgoing charged tracks with opposite charge: h^+, h^-	Topology selections
• Outgoing muon identification: 　　is_mu_prime_phast() routine • Traversed radiation lengths: 　　$X/X_0(\mu') > 30$, 　　$X/X_0(h^+, h^-) < 10$ • First/last measured track point: 　　$z_{first}(h^+, h^-, \mu') < 350\,\mathrm{cm}$, 　　$z_{last}(h^+, h^-, \mu') > 350\,\mathrm{cm}$ • Momentum determination of μ: 　　≥ 3 hits in BMS[b] • Inclusive scattering variables: 　　$Q^2 > 0.7\,(\mathrm{GeV/c})^2$, 　　$0.05 < y < 0.9$	Track ID and track quality selections
• Mass selection $M_{h^+h^-}$, 　　assuming $h^+h^- = \pi^+\pi^-$: 　　$0.5\,\mathrm{GeV/c}^2 < M_{h^+h^-} < 1.1\,\mathrm{GeV/c}^2$	ρ^0 selection
• Missing energy[c]: 　　$-4\,\mathrm{GeV} < E_{miss} < 4\,\mathrm{GeV}$ • Convergence of the kinematic fit • $\chi^2_{red,KinFit} < 10$	Exclusivity selections

[a] A primary vertex denotes a vertex which includes the beam particle. In case there is more than one primary vertex within a single event, the best primary vertex denotes the one which possesses the largest number of outgoing tracks.

[b] Beam Momentum Stations: See Sect. 3.1.

[c] The missing energy E_{miss} is given by: $E_{miss} = \frac{(p+q-\kappa)^2 - M_p^2}{2M_p}$, while M_p denotes the mass of the proton and p, q, and κ the four-momenta of the target proton, the virtual photon and the ρ^0 candidate

5.1.2 The Kinematic Fit for the Calibration of the CAMERA Detector

In order to calibrate the longitudinal position of each of the 48 counters of the CAMERA detector with respect to the spectrometer coordinate system, the longitudinal hit position of the recoiled target particle in ring A and B has to be predicted from beam and spectrometer measurements only. As one knows the position of the interaction vertex, one has to predict the polar angle of the recoiled proton. The most naive approach to calculate the polar angle makes use of the momentum balance of the reaction $\mu p \to \mu' p' \rho^0$:

$$\tan \theta_S = \frac{(\vec{p}_{p'})_T}{(\vec{p}_{p'})_L} = \frac{(\vec{p}_{\mu'} + \vec{p}_{\rho^0} - \vec{p}_\mu)_T}{(\vec{p}_{\mu'} + \vec{p}_{\rho^0} - \vec{p}_\mu)_L}. \tag{5.1}$$

The subscripts T and L denote the transverse and respectively longitudinal components of the momentum vectors. The upper left distribution of Fig. 5.1 shows the quantity θ_S as a function of the same polar angle θ_C, but using only the hit information of the CAMERA detector. It is clearly visible that there is no correlation between the two computations of the polar angle. While the quantity θ_C seems to populate the area at around 1.2 rad, the naive calculation of θ_S ends up predicting recoiled particles with

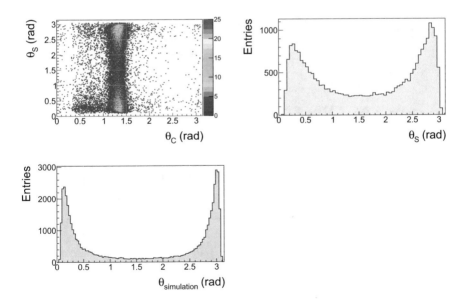

Fig. 5.1 Upper Left: The polar angle of the recoiled proton θ_S, calculated according to Eq. (5.1) as a function of the reconstructed polar angle θ_C inside the CAMERA detector. Upper Right: Projection of the upper left distribution on the θ_S axis. Bottom left: A simple toy Monte Carlo study, supposed to reproduce the behaviour of the upper right distribution

a polar angle being dominantly close to zero or in the unphysical backward scattering region.

This behaviour can be explained qualitatively. If one evaluates Eq. (5.1) within a small toy Monte Carlo study for a typical longitudinal beam and spectrometer momentum resolution of approximatelly 2 GeV/c, a mean longitudinal proton momentum of approximately 140 MeV/c and a mean transverse proton momentum of approximately 350 MeV/c, one arrives at the bottom left distribution of Fig. 5.1. These assumptions correspond to a mean polar angle of the proton of approximately 1.2 rad. Furthermore, the transverse momentum resolution was neglected for simplicity. The similarity between the top right and the bottom left distributions of Fig. 5.1 clarifies the observed behaviour and leaves the conclusion that Eq. (5.1) alone does not provide sensitivity to the polar angle of the proton.

While Eq. (5.1) makes use of the momentum balance only, it does not necessarily force the proton on its mass shell. In other words, the situation is over-constrained as the energy balance of the reaction must also be satisfied. Hence, the most clean solution of the problem is given by a kinematically constrained fit.

The measured beam and spectrometer quantities for the kinematic fit are:

$$
\vec{k} = \begin{pmatrix} k_1 \\ . \\ . \\ . \\ k_{23} \end{pmatrix} := \begin{pmatrix} \vec{a}_\mu \\ \vec{p}_\mu \\ \vec{a}_{\mu'} \\ \vec{p}_{\mu'} \\ \vec{a}_{\pi+} \\ \vec{p}_{\pi+} \\ \vec{a}_{\pi-} \\ \vec{p}_{\pi-} \\ \vec{p}_p \end{pmatrix}.
$$

The quantities $(\vec{a}_\mu, \vec{p}_\mu)^T$ and $(\vec{a}_{\mu'}, \vec{p}_{\mu'})^T$ denote the track parameters of the beam and scattered muon, the quantities $(\vec{a}_{\pi+}, \vec{p}_{\pi+})$ and $(\vec{a}_{\pi-}, \vec{p}_{\pi-})^T$ the parameters of the charged tracks. The latter are assumed to be the $\pi^+\pi^-$ pair. The target proton at rest is given by $\vec{p}_p$. Details on the definition of the track parameters and the treatment of charged tracks with respect to the kinematic fit procedure can be found in Sect. 4.3.

The unmeasured quantities are:

$$
\vec{h} = \begin{pmatrix} h_1 \\ . \\ . \\ . \\ h_6 \end{pmatrix} := \begin{pmatrix} \vec{p}_{p'} \\ \vec{v} \end{pmatrix}, \tag{5.2}
$$

where $\vec{p}_{p'}$ denotes the momentum of the outgoing proton and $\vec{v}$ is shorthand for the position of the vertex.

The kinematic fitter then calculates corrections $\Delta \vec{k}$ to the measured quantities $\vec{k}$ such that the corrected measurements:

$$\vec{k}^{fit} = \vec{k} + \Delta \vec{k},$$

together with the unmeasured quantities $\vec{h}$ minimise the least squares function of Eq. (4.1). This minimisation is performed with respect to the constraints listed in the following.

The energy and momentum conservation constraints are given, according to Sect. 4.6.1, by:

$$g_i = (p_\mu^{fit})_i - (p_{\mu'}^{fit})_i - (p_{\pi^+}^{fit})_i - (p_{\pi^-}^{fit})_i - (p_{p'}^{fit})_i = 0,$$

$$g_4 = E_\mu^{fit} + m_p c^2 - E_{\mu'}^{fit} - E_{\pi^+}^{fit} - E_{\pi^-}^{fit} - E_{p'} = 0,$$

$\forall\, i \in \{1, 2, 3\}$, while the index denotes Cartesian components of the three-vectors.

The variables denoted with the superscript "fit" emphasise the fact that the quantities corrected by the kinematic fit have to satisfy the constraints. Apart from the energy and momentum conservation all tracks except the initial and final state proton must join in a common vertex:

$$g_{4+i} = (p_\mu^{fit})_3 \left(v_i - (a_\mu^{fit})_i \right) - (p_\mu^{fit})_i \left(v_3 - (a_\mu^{fit})_3 \right) = 0,$$

$$g_{6+i} = (p_{\mu'}^{fit})_3 \left(v_i - (a_{\mu'}^{fit})_i \right) - (p_{\mu'}^{fit})_i \left(v_3 - (a_{\mu'}^{fit})_3 \right) = 0,$$

$$g_{8+i} = (p_{\pi^+}^{fit})_3 \left(v_i - (a_{\pi^+}^{fit})_i \right) - (p_{\pi^+}^{fit})_i \left(v_3 - (a_{\pi^+}^{fit})_3 \right) = 0,$$

$$g_{10+i} = (p_{\pi^-}^{fit})_3 \left(v_i - (a_{\pi^-}^{fit})_i \right) - (p_{\pi^-}^{fit})_i \left(v_3 - (a_{\pi^-}^{fit})_3 \right) = 0,$$

$\forall\, i \in \{1, 2\}$, while the index denotes Cartesian components of the three-vectors.

For each charged track two vertex constraints are entering the system of equations. They are explained in Sect. 4.6.2. The initial and final state proton are not constrained to the vertex. Since for either of them there is no measurement of their two transverse coordinates at a certain longitudinal position, one would have to introduce two free parameters, which are then trivially fixed by the two additional vertex constraints. This "zero-sum game" of adding two constraints, while at the same time being obliged to include two free parameters, is not "played".

In total 12 constraints are introduced into the procedure, while according to Eq. (5.2) six free parameters have to be determined. Hence, the number of degrees of freedom is six.

In this context the most important feature of the kinematic fit is that it provides a precise determination of the polar angle of the recoiled particle, using beam and spectrometer quantities only. This allows for a calibration of the longitudinal positions of the scintillating counters. The calibration procedure of the longitudinal positions of the 48 scintillating counters is described in Sect. 5.1.4.

5.1.3 Calibration of the Azimuthal Angle

In order to calibrate the azimuthal position of the 48 scintillating counters of ring A and ring B with respect to the spectrometer coordinate system, the azimuthal angle ϕ of the recoiled particle determined by the kinematic fit is used. This determination of ϕ relies solely on the measurement of the beam and scattered muon tracks together with the two charged pion tracks and the assumption of exclusivity. It is independent of the collected hit information of the CAMERA detector.

Within the first step of the calibration procedure the distribution of the ϕ angle is separated for each of the 48 counters of the CAMERA detector. A distribution similar to the right side of Fig. 5.2 or Fig. 5.3 results. For each reconstructed hit in a

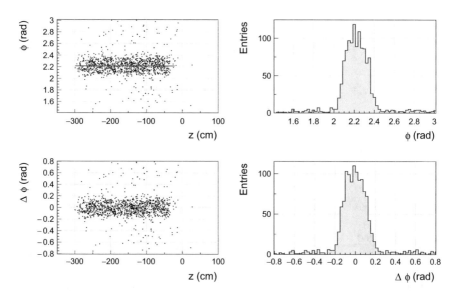

Fig. 5.2 Upper left: Azimuthal angle ϕ as a function of the reconstructed z-position of the hits detected inside the B0 counter of CAMERA. The red line indicates a function of the form $\phi(z) = \Phi_{B0} = const$. The value of the extracted parameter Φ_{B0} gives the azimuthal position of the counter B0 with respect to the spectrometer coordinate system. Upper right: Projection of the upper left distribution on the ϕ-axis. Lower left: Distribution of $\Delta\phi = \phi - \phi_C$ as a function of z. Lower right: Projection of the lower left distribution on the $\Delta\phi$-axis. The variable ϕ denotes the azimuthal angle of the recoiled particle, determined according to Sect. 5.1.2 by spectrometer measurements only, while ϕ_C denotes the azimuthal angle of the counter B0, using the calibration values indicated by the red line

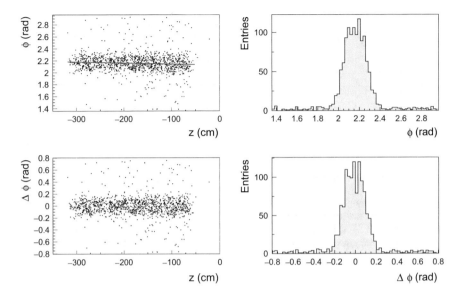

Fig. 5.3 Upper left: Azimuthal angle ϕ as a function of the reconstructed z-position of the hits detected inside the A0 counter of CAMERA. The red line indicates a function of the form $\phi(z) = m_{A0}z + \Phi_{A0}$. The values of the extracted parameters m_{A0} and Φ_{A0} give the azimuthal position of the counter A0 as a function of the reconstructed z-position with respect to the spectrometer coordinate system. Upper right: Projection of the upper left distribution on the ϕ-axis. Lower left: Distribution of $\Delta\phi = \phi - \phi_C$ as a function of z. Lower left: Projection of the lower left distribution on the $\Delta\phi$-axis. The variable ϕ denotes the azimuthal angle of the recoiled particle determined by spectrometer measurements only according to Sect. 5.1.2, while ϕ_C denotes the azimuthal angle of the counter A0 using the calibration indicated by the red line

certain counter the determined ϕ angle enters the respective ϕ-distribution associated to this counter. The central values of these distributions correspond to the azimuthal positions of the counters and are regarded as calibration constants of the first iteration. Using these constants, the background in the sample can be further suppressed by applying a soft cut on coplanarity according to a distribution similar to Fig. 5.4. This allows for a first calibration of the longitudinal hit position, according to Sect. 5.1.4. Furthermore, by analysing the width of the rectangular shaped ϕ distributions a transverse displacement of ring A and B with respect to the transverse origin is observed. The detailed analysis of the ϕ distributions can be found within Ref. [2].

The second step of the calibration is to replace the cut on the coplanarity, useful in case of the longitudinal hit position calibration, with a soft vertex pointing cut, according to Fig. 5.9. Next, the value of ϕ is correlated with the reconstructed z-position inside the counters for each of the counters individually. The two dimensional distributions, illustrating the correlation of ϕ and z, are shown on the left side of Figs. 5.2 and 5.3 for ring B and A. In case of ring B no correlation between the two quantities is observed and the single calibration constant Φ_{Bi} represents the azimuthal position of the ith counter. The calibrated ϕ angle of a recoiling particle traversing a ring B element i is thus given by:

$$\phi_i(z) = \Phi_{Bi} = const \quad i \in \{0, \ldots, 23\}.$$

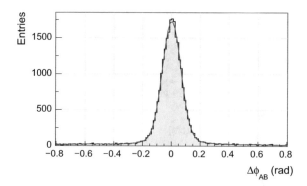

Fig. 5.4 The distribution of $\Delta\phi_{AB} = \phi_{AB} - \phi$, indicating an upper limit for the resolution of the azimuthal angle of the recoiled particle achievable by CAMERA. The quantity ϕ_{AB} is calculated according to Eq. (5.3), while ϕ denotes the azimuthal angle of the recoiled particle determined by spectrometer measurements only according to Sect. 5.1.2. The full width at half the maximum value is given by $\sigma_{FWHM} = 8.9°$

In case of ring A a correlation between ϕ and z is observed and parametrised by a linear function. It is related to a slight twist of the thin and deformable scintillating counters of ring A, introduced during its assembly. The calibrated ϕ angle of a recoiling particle traversing the ith element of ring A is thus given by the two calibration constants m_{Ai} and Φ_{Ai}, according to:

$$\phi_i(z) = m_{Ai} z + \Phi_{Ai} \quad i \in \{0, \ldots, 23\}.$$

The functional form of the parametrisations is indicated on the top left side of Figs. 5.2 and 5.3 by the red lines. At the bottom of these figures the difference $\Delta\phi$ of the azimuthal angle given by the kinematically constrained fit and the azimuthal angle measured by CAMERA is shown after the application of the calibration constants. These distributions should serve as a visual proof of concept. Figures A.2 and A.1 inside Appendix A.1.1 show the top right distribution of Figs. 5.2 and 5.3 for all 48 counters.

Since the counters of ring A are displaced by 7.5° with respect to the counters of ring B, the azimuthal resolution of a recoiling particle, can be increased by using the following definition of the azimuthal angle:

$$\phi_{AB} = \frac{\phi_A + \phi_B}{2}. \tag{5.3}$$

Fig. 5.4 shows the achievable azimuthal resolution of CAMERA with respect to the spectrometer and beam measurements. The full width at half the maximum value is given by:

$$\sigma_{FWHM} = 8.9°.$$

The bare full azimuthal width of a counter corresponds to $\frac{360°}{24} = 15°$. By using Eq. (5.3) a full azimuthal width of 7.5° degrees is expected in case one would assume a negligible ϕ resolution of the beam and spectrometer measurement. However, a negligible beam and spectrometer resolution is already ruled out by the non rectangular shape of the distribution.

5.1.4 Calibration of the Longitudinal Position

As described in Sect. 5.1.2, the calibration of the longitudinal hit positions of the 48 scintillating counters of CAMERA with respect to the COMPASS coordinate system strongly relies on the kinematically constrained fit.

In case of the calibration of the ring B elements the position of the interaction vertex $\vec{v}$, the polar angle θ of the recoiled proton and the distance of the ring B element with respect to the z-axis r_B are used. The longitudinal hit position can be predicted as illustrated in Fig. 5.5. The z-position z'_B, with respect to the interaction vertex, at which the recoiling particle intersects ring B is calculated as follows:

$$z'_B = \frac{r'_B}{\tan(\theta)},$$

while the quantity:

$$r'_B = r_B - d = r_B - \sqrt{v_x^2 + v_y^2},$$

denotes the transverse distance between the interaction vertex and ring B. The absolute z-position, with respect to the COMPASS coordinate system, follows as:

$$z_B = v_z + z'_B.$$

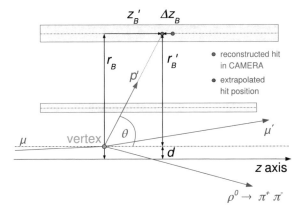

Fig. 5.5 Schematic illustration of the longitudinal hit position calibration of the ring B elements

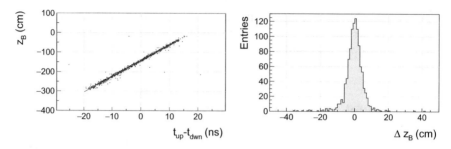

Fig. 5.6 Left: Distribution of the predicted z-position z_B of the recoiled particle inside the B0 counter of CAMERA as a function of $(t_u - t_d)$, the difference of the up- and downstream time-stamps measured with the two photomultiplier tubes of the B0 counter. According to Eq. (3.1) of Sect. 3 a function of the form $z_{B0}(\Delta t) = \frac{1}{2}c_{B0}(t^u_{B0} - t^d_{B0}) + k^z_{B0}$ is shown in red, which illustrates the determination of the calibration parameters c_{B0} and k^z_{B0}. Right: Distribution of $\Delta z_B = z_B - z_{C,B}$, the difference between the predicted z-position z_B, using the kinematically constrained fit, and the reconstructed z-position $z_{C,B}$ after the application of the calibration constants c_{B0} and k^z_{B0}

The left distribution of Fig. 5.6 shows z_B as a function of the difference between the up- and downstream time-stamps $(t_u - t_d)$, measured at the two sides of an exemplary ring B element. The slope $\frac{1}{2}c_{Bi}$ and the offset k^z_{Bi} of the distribution are extracted as indicated by the red line. Hence, the z-position of a particle traversing the i th element of ring B is given according to Eq. (3.1) by:

$$z_{C,Bi} = \frac{1}{2}c_{Bi}(t^u_{Bi} - t^d_{Bi}) + k^z_{Bi} \quad , i \in \{0, \ldots, 23\}. \tag{5.4}$$

To make use of the good position resolution of the ring B elements, a slightly different approach is chosen for the longitudinal hit position calibration of ring A. The hit position inside ring A is calculated by an interpolation between the interaction vertex and $z'_{C,B}$, the reconstructed hit position in ring B with respect to the interaction vertex. As illustrated in Fig. 5.7, the z-position z'_A at which the recoiling particle intersects ring A, is calculated with respect to the interaction vertex as follows:

$$z'_A = \frac{r'_A}{r'_B}z'_{C,B},$$

while the quantity:

$$r'_A = r_A - d = r_A - \sqrt{v_x^2 + v_y^2},$$

denotes the transverse distance between the interaction vertex and ring A. The absolute z-position, with respect to the COMPASS coordinate system, follows as:

$$z_A = v_z + z'_A.$$

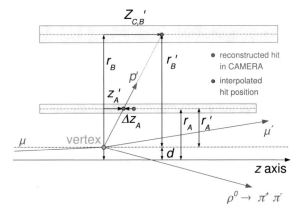

Fig. 5.7 Schematic illustration of the longitudinal hit position calibration of the ring A elements

As in case of the ring B calibration Fig. 5.8 shows z_A as a function of the difference between the up- and downstream time-stamps $(t_u - t_d)$, measured at the two sides of an exemplary ring A element. The z-position of a particle traversing the ith element of ring A is thus given according to Eq. (3.1) by:

$$z_{C,Ai} = \frac{1}{2}c_{Ai}(t_{Ai}^u - t_{Ai}^d) + k_{Ai}^z \quad , i \in \{0, \ldots, 23\}. \tag{5.5}$$

The distributions of the left side of Figs. 5.6 and 5.8 for each of the 48 scintillating counters are shown in Appendix A.1.2 inside Figs. A.3 and A.4. The right sides of Figs. 5.6 and 5.8 show the distributions of the difference Δz_B and Δz_A for an examplaric ring B and A counter. As illustrated in Figs. 5.5 and 5.7, the quantities Δz_B and Δz_A represent the difference between the predicted and the reconstructed

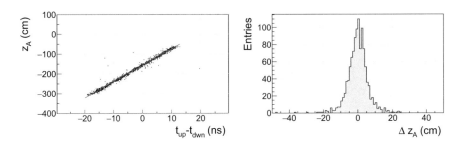

Fig. 5.8 Left: Distribution of the predicted z-position of the recoiled particle inside the A0 counter of CAMERA as a function of $(t_u - t_d)$, the difference of the up- and downstream time-stamps measured with the two photomultiplier tubes of the A0 counter. According to Eq. (3.1) of Sect. 3 a function of the form $z_{A0}(\Delta t) = \frac{1}{2}c_{A0}(t_{A0}^u - t_{A0}^d) + k_{A0}^z$ is shown in red, which illustrates the determination of the calibration parameters c_{A0} and k_{A0}^z. Right: Distribution of $\Delta z_A = z_A - z_{C,A}$, the difference between the predicted z-position z_A, using an interpolation between the interaction vertex and the hit position in ring B, and the reconstructed z-position $z_{C,A}$ after the application of the calibration constants c_{A0} and k_{A0}^z

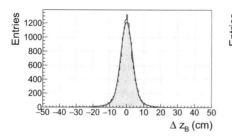
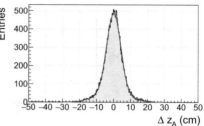

Fig. 5.9 Distributions of the difference $\Delta z_{A;B} = z_{A;B} - z_{C,A;B}$ between the predicted z-position $z_{A,B}$ of the recoiled particle and the reconstructed z-position $z_{C,A;B}$ inside CAMERA after application of the respective calibration constants. All counters of ring B, respectively ring A, of the CAMERA detector are shown inside these distributions. Left: The quantity $z_{C,B}$ is calculated using the kinematically constrained fit described in Sect. 5.1.2. Right: The quantity $z_{C,A}$ is calculated using an interpolation between the interaction vertex and the hit position in ring B. The red lines show Gaussian fits applied to the distributions

longitudinal hit positions. As one accumulates these distributions for all scintillating counters of ring B and A, the left and right side of Fig. 5.9 result.

From Fig. 5.9 an upper limit on the position resolution of ring B and A is determined by a Gaussian fit. In case of ring B this results in:

$$\sigma_B = 3.3\,\text{cm}, \tag{5.6}$$

while in case of ring A a value of:

$$\sigma_A = 4.1\,\text{cm}, \tag{5.7}$$

is extracted. It should be emphasised that these values are extracted with respect to the θ and vertex resolution in case of ring B and with respect to the vertex and ring B resolution in case of ring A. They do not reflect the bare resolutions of the counters. Nevertheless these resolutions serve as a starting point for the kinematic fitting procedure and for the simulations, as it is explained in detail in Sects. 4.5 and 6.1.3.

5.1.5 Momentum Calibration

After the calibration of the azimuthal positions of the scintillating counters and the longitudinal hit position within the counters, the distance of flight of a particle traversing ring A and B of CAMERA is determined by Eq. (3.2). Hence, a calibration of the time of flight, given by Eq. (3.3), can be attacked.

As illustrated in Fig. 3.7, one counter of ring A corresponds to two counters of ring B, which results from the azimuthal shift of 7.5 degrees between the inner and outer ring of CAMERA. Thus, in order to calibrate the time of flight, 48 calibration constants:

$$k_{lm}^T \text{ with } l \in \{0, \ldots, 23\} \text{ and } m \in \{l, (l+1) \bmod 24\},$$

have to be determined. In analogy to Eq. (3.3) the calibrated time of flight for the counters B_l and A_m is then explicitly given by:

$$T_{lm} = \frac{t_l^u + t_l^d}{2} - \frac{t_m^u + t_m^d}{2} + k_{lm}^T.$$

Due to instabilities within the readout electronics during the 2012 run, these constants had to be determined on a run by run basis. The method relies on particles traversing CAMERA with the speed of light. Using the value of the speed of light and the reconstructed distance of flight, the time of flight of these particles can be predicted. A comparison with the reconstructed time of flight yields the offsets k_{lm}^T. The detailed procedure is described in Ref. [2]. It includes run by run stability checks, which show a clear correlation with clock instabilities caused by firmware reloads of the CAMERA readout electronics.

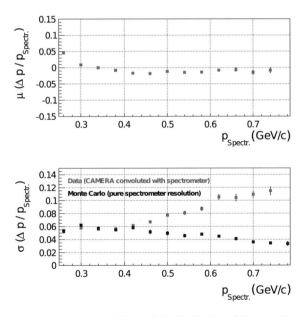

Fig. 5.10 The mean value μ and the width σ of the distribution of the quantity $(p_C - p_{Spectr.})$. The quantity p_C denotes the magnitude of the reconstructed momentum using CAMERA, the quantity $p_{Spectr.}$ the magnitude of the momentum of the recoiled target proton determined within the exclusive ρ^0 sample and the hypothesis of exclusivity from beam and spectrometer measurements only. The data is shown in red and yields the convoluted momentum resolution of CAMERA and the beam and spectrometer prediction. A ρ^0 Monte Carlo sample is shown in black, from which the momentum resolution of a pure Beam and Spectrometer measurement is estimated

The momentum between ring A and B is determined via Eq. (3.5) from the extracted set of constants k_{lm}^T. In order to translate the momentum between ring A and B to the momentum at the interaction vertex, energy loss corrections of the proton within the target and ring A have to be applied. These corrections are calculated as described in Ref. [3] and are reviewed in Ref. [2].

Finally, the reconstructed proton momentum at the interaction vertex determined by CAMERA is compared to the proton momentum predicted by the kinematic fit within the exclusive ρ^0 sample. The comparison is shown in Fig. 5.10 and illustrates that the momentum is well calibrated.

Within the upper plot of Fig. 5.10 one observes a slight bias appearing at momenta below 0.28 GeV/c. As one translates this momentum to the square of the four momentum transfer to the proton, it corresponds to a value of $|t|$ below 0.08 (GeV/c)^2, which is the lower bound on the extraction region of the DVCS cross section (see Chap. 7). Thus, the bias is not further taken into account. From the lower part of Fig. 5.10 it is evident that the measured resolution is almost purely given by the beam and spectrometer measurement at small momenta, while at large momenta the extracted resolution is governed by the CAMERA detector. This emphasis why a kinematically constrained fit including both the CAMERA detector and the spectrometer will yield the best possible resolution for the momentum of the recoiled proton. The combination of the beam, spectrometer and CAMERA measurements within a kinematically constrained fit will be performed within the analysis of exclusive single photon production, discussed from Chap. 6 onwards.

5.2 Luminosity Determination

The integrated luminosity $\mathcal{L}^\pm$ for the μ^+ and the μ^- data yield, denoted by $\pm$, is calculated according to Eq. (5.8):

$$\mathcal{L}^\pm = \frac{\rho_{\mathrm{LH_2}} N_a l}{m_p} \Phi_{eff}^\pm. \tag{5.8}$$

The quantity $m_p = 1.0078 \, \frac{\mathrm{g}}{\mathrm{mol}}$ denotes the molar proton mass, the quantity $\rho_{\mathrm{LH_2}} = 0.0704 \, \frac{\mathrm{g}}{\mathrm{cm^3}}$ the density of protons inside the liquid hydrogen target, $N_a = 6.022 \cdot 10^{23} \, \frac{1}{\mathrm{mol}}$ the Avogadro constant, $l = 240 \, \mathrm{cm}$ the effective target length and $\Phi_{eff}^\pm$ the total effective number of muons traversing the target during the μ^+ or respectively the μ^- data taking periods. In order to determine the quantity $\phi_{eff}^\pm$ two different methods have been used and cross checked amongst each other.

Both methods require the same definition of a beam track. In order to have a precise measurement by the beam telescope, a "good" beam track is required to have at least two hits in the scintillating fibre detectors and at least three hits in the silicon detectors. Furthermore, since the analysis of exclusive reactions demands a precise determination of the beam momentum, it is required to observe at least three hits in the beam momentum stations. Finally, the track has to traverse the full target, as it will be described in Sect. 6.2.1. In order to make use of the flux values in the analysis of exclusive reactions, exactly the same conditions have to be applied in the corresponding event selections.

The first method relies on the analysis of random trigger events. The number of beam tracks during a spill is counted for pure random trigger events. A beam track is identified in case the time of the beam particle, measured by the beam telescope, is compatible with the time of the trigger within a time window of $\pm 8\,\text{ns}$. The number of beam tracks within this time window of $16\,\text{ns}$ is then extrapolated to the total duration of a spill, which yields the flux Φ. It shall be emphasised that in contrast to the physical triggers, the random trigger is not connected to the beam veto system. Thus, a correction taking into account the Veto Dead Time c_{vdt}, according to Eq. (5.9), must be applied. This is necessary in order to compare the results with the second method and to calculate the luminosity for the extraction of a cross section:

$$\Phi_{eff} = (1 - c_{vdt})\Phi. \tag{5.9}$$

The veto lifetime, $c_{vlt} = (1 - c_{vdt})$, describes the probability that the trigger signal is not coincidentally suppressed by the veto system. It is determined by the fraction of trigger attempts for which a time shifted veto is applied in coincidence with the physics-trigger signal and the overall number of trigger attempts. For the overall number of trigger attempts no veto has been applied [4]. The shift in time is usually around 35–$40\,\text{ns}$ and ensures that the veto and the trigger signal are uncorrelated. For the 2012 data taking the time period, for which the attempts are counted, is chosen to be the duration of the spill and the correction for the veto dead time has been taken into account on a spill by spill basis.

The second method relies on the good knowledge of the structure function F_2^p. The integrated luminosity is calculated according to Eq. (5.10):

$$\mathcal{L} = \frac{1}{\sigma_\Omega} \sum_{i=1}^{N_{meas} \in \Omega} \frac{\eta(Q_i^2, (x_{Bj})_i)}{A(Q_i^2, (x_{Bj})_i)}, \tag{5.10}$$

while σ_Ω, the integrated differential cross section over the phase space element Ω, is given by [5]:

$$\sigma_\Omega = \int_\Omega \frac{4\pi\alpha^2}{Q^4} \frac{F_2^p(x_{Bj}, Q^2)}{x_{Bj}} \left(1 - y - \frac{Q^2}{E_l^2} + \left(1 - \frac{2m_l^2}{Q^2}\right) \frac{y^2 + Q^2}{2E_l^2(1 + R(x_{Bj}, Q^2))}\right) dx_{Bj} dQ^2.$$

Hence, a typical measurement of F_2^p is reversed. The luminosity is calculated from the known values of F_2^p, given by the Tulay's fit [6], R the ratio of the longitudinal and transverse cross sections [7] and the experimentally measured number of scattered muons N_{meas} into the phase space element Ω, taking into account radiative corrections $\eta(Q^2, x_{Bj})$ and experimental acceptance $A(Q^2, x_{Bj})$. The detailed procedure of the second method is described in Ref. [8].

Using the second method together with Eq. (5.8), the number of muons traversing the target during each spill of the 2012 data taking can be extracted. It is compared

Table 5.2 Integrated muon flux Φ, integrated effective muon flux Φ_{eff} and the integrated luminosity $\mathcal{L}$ for the 2012 data taking period. The values have been calculated according to Sect. 5.2 and the stability criteria of Sect. 5.3 are taken into account

Data yield	μ^+ and μ^-	μ^+	μ^-
$\Phi/10^{12}$	5.203	2.582	2.622
$\Phi_{eff}/10^{12}$	4.201	1.871	2.330
$\mathcal{L}/(\mathrm{pb}^{-1})$	42.38	18.88	23.51

with the result given by the first method, using Eq. (5.9). In Ref. [8] it is shown that the results of the two methods agree well. The statistical error of the second method exceeds the one of the first method. Hence, the first method is used for the extraction of the DVCS cross section and it is decided to consider a systematic effect of three percent on the flux determination. The lists containing the flux and the veto dead time correction for each spill have been provided by Ref. [9].

Table 5.2 shows the integrated muon flux Φ, the integrated effective muon flux Φ_{eff} and the integrated luminosity $\mathcal{L}$ for the 2012 data taking period. The stability criteria of Sect. 5.3 are taken into account.

5.3   Data Quality

Several spill by spill stability checks have been applied to the data (see Table 5.3). They can be divided into six categories:

- **Spectrometer stability**:
  Studying several meaningful variables, like the number of tracks per primary vertex[1] or the number of primary vertices per event as a function of the spill number, suspicious spills are excluded from the physics analysis. The decision, if a spill is excluded, is based on the number of neighbouring spills. Neighbouring spills are defined as spills which show values for the investigated set of variables compatible within a window of several standard deviations with respect to the values of the investigated spill. The number of standard deviations and the number of required neighbouring spills in order to classify a spill as good or bad depends on the set of variables. A detailed description of the method can be found in Ref. [10], while for the actual production of the bad spill list for 2012 it shall be referred to Ref. [11].
- **Internal synchronisation of the CAMERA readout**:
  The synchronisation state of the operating clock of the CAMERA readout with respect to the clock provided by the trigger control system was monitored continuously during the 2012 data taking. In case the synchronisation was lost the corresponding spills were excluded from the analysis.

---

[1] A primary vertex denotes a vertex which includes the beam particle.

Table 5.3 Lost percentage of the accumulated muon flux after a successive application of the stability criteria. The numbers of each column include the application of the criteria above the respective column

Stability criterion	All data (%)	μ^+ data (%)	μ^- data (%)
Time of Flight calibration for the CAMERA detector	0.6	0.9	0.2
Internal synchronisation of the CAMERA readout	1.8	2.9	0.6
Hit rate stability for the CAMERA detector	2.7	4.1	1.0
Time synchronisation of fibre station 2 readout	5.1	7.5	2.3
Spectrometer stability	9.3	12.1	6.1
Synchronisation of the CAMERA readout to the trigger signal	15.6	22.0	8.2

- **Time of Flight calibration for the CAMERA detector**:
  As mentioned in Sect. 5.1.5, the time of flight calibration of the CAMERA detector is performed on a run by run basis. For certain runs the time of flight calibration constants were found to change within a single run. In these cases the spills were omitted. The detailed procedure is explained in Ref. [2].
- **Synchronisation of the CAMERA readout to the trigger signal**:
  In order to provide a time measurement within CAMERA with respect to the trigger signal, the time of the trigger signal is measured by the so called Master Time clock. At the beginning of a data recording phase the time measurement of the CAMERA readout has to be synchronised with the time measurement of the Master Time clock. It was observed that in certain cases this synchronisation failed and the corresponding spills were excluded from the physics analysis.
- **Hit rate stability for the CAMERA detector**:
  The number of hits observed in ring A and ring B normalised to the flux are checked for each spill. Certain spills, differing largely from the average, were excluded from the measurement, as it can be seen in Appendix A.3 inside Fig. A.25.
- **Time synchronisation of fibre station 2 readout**: During the 2012 data taking the readout of fibre station 2 was performed with the M1 TDC [12]. The synchronisation of the M1 readout electronics with the trigger control system was done spill by spill. For certain spills time jumps of the measured hit time of fibre station 2 with respect to the trigger signal were observed and excluded from the measurement [13].

Furthermore, an additional problem with the CAMERA readout in 2012 was observed. It is related to the sampling of the ADC information. Each bit of the digitised analog signals of the 96 photomultipliers is transmitted from the ADCs to the GANDALF main FPGA. After the data taking sticky bits and random bit-flips were identified. They are related to the initialisation state of the FPGA as it is explained in Sect. 9.2.2. Thus, certain elements of the CAMERA detector are excluded from the analysis for certain runs. The method to detect these runs is based on a Fourier transformation of the up- and downstream time difference spectra of the scintillating counters. It is described in detail in Ref. [2]. In order to account for this data loss in the analysis and to prevent an azimuthal systematic distortion, the effect is introduced into the simulations as follows.

For each pair of corresponding ring A and ring B scintillators (i, j) a probability $p(i, j)$ is calculated, according to Eq. (5.11):

$$p(i, j)^{\pm} = \frac{\sum_{m \in M^{\pm}} \sum_{l \in L_{ij}^{\pm}} \Phi_l \delta_{lm}}{\sum_{m \in M^{\pm}} \Phi_m} = \frac{\sum_{l \in L_{ij}^{\pm}} \Phi_l}{\sum_{m \in M^{\pm}} \Phi_m}. \tag{5.11}$$

The quantity $M^{\pm}$ denotes the set of all run numbers considered in the analysis for the corresponding beam charge $\pm$, while $L_{ij}^{\pm}$ denotes the runs for which the segment (i, j) was operational for the corresponding beam charge $\pm$. The quantity Φ_m denotes the total number of muons, which traversed the target during the run number m. The Kronecker delta is depicted by δ. Inside the Monte Carlo simulations the percentage $(1 - p(i, j)^{\pm})$ of the data for the segment (i, j) is rejected. Table A.1 in Appendix A.3 shows the calculated values for $p(i, j)$ for each segment. Assuming no azimuthal systematic effects, the overall data loss due to the "bit-flip issue" is approximately 11%. This value agrees for the overall data taken with the μ^+ and the μ^- beam on the level of 0.5%.

5.4   Determination of the Efficiency of CAMERA

The technique to determine the efficiency of the CAMERA detector relies on the relation:

$$E = \frac{N_S}{N_O}. \tag{5.12}$$

The quantity E denotes the efficiency, N_O the number of observable events and N_S the number of observed events within N_O. The idea is to produce a data sample N_O by using the exclusive ρ^0 channel in order to provide an efficiency separately for each side of the scintillating counters of the CAMERA detector. This is achieved by checking which fraction of the observable events N_O is actually observed by either side of the counter. To minimise the amount of background events within the N_O sample, the full hit information within the recoil detector apart from the information of the side which is under investigation is used.

In the following the details of the procedure are explained, using the upstream side determination of the efficiency of ring A as an example. In this case the beam and spectrometer measurements together with the hit information observed in ring B and at the downstream side of ring A are combined to build the N_O sample. The generation of the N_O sample can be split into two stages:

The first stage combines the measured time-stamps t_B^u, t_B^d and the measured signal amplitudes A_B^u, A_B^d, given at the up-(u) and downstream (d) side of ring B. The momentum vector of the proton deduced by the kinematically constrained fit together with the interaction vertex is used to pinpoint the expected azimuthal angle and z-position z_B^P of the recoiled proton within ring B. Using the prediction of the proton azimuth, the corresponding counter is selected. Next, a sharp cut on the distribution of the difference between z_B^P and z_B is performed. The latter quantity denotes the z-position in ring B reconstructed by the measured time-stamps t_B^u and t_B^d. The distribution and the applied cut are shown on the left side of Fig. 5.11. In addition, the inter-calibration of one of the fibre stations (start counter), comprising the time measurements of the incoming muon and ring B [2], is exploited. It yields the magnitude of the proton momentum denoted as p_B^{SC}. The quantity p_B^{SC} is compared with p_F, the magnitude of the proton momentum predicted by the kinematic fit. The distribution of $(p_F - p_B^{SC})$ and the applied cut are shown on the right side of Fig. 5.11. Finally, $\sqrt{A_B^u A_B^d}$ as a function of p_F results in the distribution shown in Fig. 5.12 and allows for a further cut on the proton signature.

The second stage uses the information given by ring B together with measured time stamps t_A^d and the measured signal amplitudes A_A^d of the downstream (d) side of ring A. Since two A counters correspond to one B counter, the selection of the A counter is again based on the proton azimuth predicted by the kinematic fit. To determine the predicted z-position z_A^P within ring A, an interpolation between the measured hit within ring B together with the interaction vertex is used. Solving Eq. (3.1) for t_A^u, using the measured value for t_A^d and $z_A = z_A^P$, the time-stamp of the upstream side

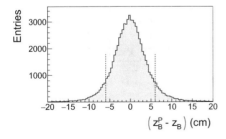
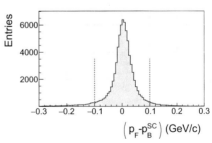

Fig. 5.11 Left: Distribution of the difference between the predicted z-position z_B^P in ring B using the kinematically constrained fit and the reconstructed z-position z_B determined by the up and down time-stamps of ring B. Right: Distribution of the difference between the proton momentum deduced with the kinematically constrained fit and the proton momentum reconstructed with the time-stamps of ring B and the time-stamp of the incoming muon passing the startcounter. The blue lines indicate the cuts applied in order to select the N_O sample, corresponding to Eq. (5.12)

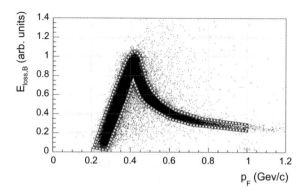

Fig. 5.12 Distribution of the energy loss in ring B, $E_{loss,B} = \sqrt{A_B^u A_B^d}$, as a function of the proton momentum p_F deduced with the kinematically constrained fit. The blue polygon indicates the cut applied in order to select the N_O sample, corresponding to Eq. (5.12)

is predicted. This time-stamp shall be denoted as $t_A^{u,P}$ in the following, to underline the fact that it is not a directly measured quantity. It is used within Eqs. (3.3) and (3.5) to determine the proton momentum p_C. The distributions of $(p_B^{SC} - p_C)$ and $(p_F - p_C)$ are shown in Fig. 5.13 together with the applied cuts. In addition, using the amplitude A_A^d together with the hit position z_A^P, the signal amplitude at the upstream side can be predicted by the relation:

$$A_A^{u,P} = A_A^d \exp\left(\frac{-(z_A^P - z_E^u)}{L}\right), \tag{5.13}$$

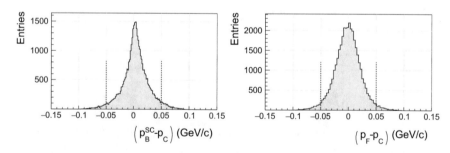

Fig. 5.13 Left: Distribution of the difference between the proton momentum p_B^{SC}, deduced with the startcounter and ring B, and the proton momentum p_C, deduced with the time-stamps of ring B together with the z-position and the downstream time-stamp of ring A. Right: Distribution of the difference between the proton momentum p_F given by the kinematically constrained fit and the proton momentum p_C. The blue lines indicate the cuts applied in order to select the N_O sample, corresponding to Eq. (5.12)

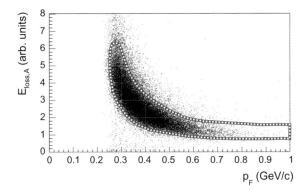

Fig. 5.14 Distribution of the energy loss $E_{loss,A} = \sqrt{A_A^{u,P} A_A^d}$ as a function of p_F, the proton momentum deduced with the kinematically constrained fit. The quantity $A_A^{u,P}$ denotes the predicted upstream amplitude in ring A, given by Eq. (5.13), while A_A^d denotes the measured downstream amplitude in ring A. The blue polygon indicates the applied cut in order to select the N_O sample, corresponding to Eq. (5.12)

while L is the attenuation length of the counter and z_E^u the z-position of the upstream end of the counter. The measured downstream amplitude A_A^d, the predicted amplitude at the upstream end $A_A^{u,P}$ and p_F are combined to procude the distribution shown in Fig. 5.14. It is also used to further suppress the background within the N_O sample, according to the indicated two dimensional cut.

So far, no measured information of the upstream side of ring A has been used. In order to build the N_S sample, the time-stamp t_A^u is checked for compatibility with $t_A^{u,P}$. Figure 5.15 shows the distribution of $(t_A^u - t_A^{u,P})$. The corresponding distributions of several bins in the longitudinal position are shown in Appendix A.2.1 inside Fig. A.5. In case of the downstream side efficiency determination of ring A the procedure is analogue. The corresponding distributions are shown in Appendix A.2.1 inside Figs. A.9–A.12.

For the determination of the ring B efficiency the procedure is almost analogue. Due to the high background rate in ring A, no inter-calibration between ring A and the startcounter was produced. Thus, no counter parts to the right side of Fig. 5.11 and the left side of Fig. 5.13 exist. The corresponding distributions are shown in Appendix A.2.2 inside Fig. A.14 for the selection of the N_O sample using ring A, inside Figs. A.15–A.19 for the upstream side and inside Figs. A.20–A.24 for the downstream side efficiency determination of ring B.

Table 5.4 lists the efficiency of ring A for the different bins in t, the square of the four-momentum transfer to the proton, which are later used in Sect. 7. No noticeable trend is observed for the efficiency as a function of t. Thus, it is decided to parametrise the efficiency for the up- and downstream side of ring A as a function of the z-position of the recoiled particle inside the scintillators of ring A.

Figure 5.16 shows the dependence of the efficiency on the z-position. It is clearly visible that the efficiency decreases with increasing distance of the hit position to the ends of the scintillators. This is a direct result of low photo-electron statistics in case

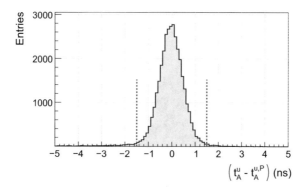

Fig. 5.15 Distribution of the difference between the measured upstream time-stamp t_A^u and the predicted time-stamp $t_A^{u,P}$, deduced using Eq. (3.1) together with the measured downstream time-stamp and the expected z-position in ring A. The blue line indicates the applied cut in order to select the N_S sample, corresponding to Eq. (5.12). It should be underlined that this cut is the first one, which is supposed to select the full signal in order to not artificially decrease the efficiency. The corresponding distributions of several bins in the longitudinal hit position are shown in Fig. A.5 and separated for the data yields taken with the μ^- and μ^+ beam inside Figs. A.6 and A.7

Table 5.4 Ring A efficiencies as a function of t, the square of the four-momentum transfer to the proton

$|t|$ range in (GeV/c)^2	Upstream efficiency	Downstream efficiency	Ring A efficiency
]0.08, 0.22]	0.956 ± 0.001	0.930 ± 0.001	0.889 ± 0.002
]0.22, 0.36]	0.955 ± 0.003	0.928 ± 0.003	0.886 ± 0.005
]0.36, 0.5]	0.941 ± 0.008	0.946 ± 0.009	0.89 ± 0.01
]0.5, 0.64]	0.95 ± 0.02	0.92 ± 0.02	0.87 ± 0.03

of the thin ring A elements. The efficiency drop at the far end is more pronounced in case of the downstream side, for which the light guide is significantly longer compared to the upstream side.

For the upstream side efficiency a drop close to the upstream end of ring A is observed. This is related to the high voltage setting of the photomultipliers, for which a certain trade-off had to be found during the hardware commissioning procedure. Setting the high voltage too low, causes a loss of signals at the far end, while setting the voltage too high, causes the signals at the near end to exceed the dynamic range of the electronics. In case the dynamic range of the electronics is exceeded, a clipping of the signals is observed. This causes an unpredictable distortion of the time-stamps of the processed photomultiplier signals and results in a decreasing efficiency. Looking at Fig. 5.17 one can observe that this drop is more severe for the scintillators marked as "bad quality".

During the assembly of ring A each scintillator was tested for its attenuation length and it was observed that certain scintillators possess much smaller values for

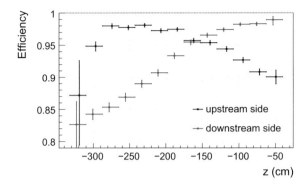

Fig. 5.16 The efficiency of ring A as a function of the z-position, given by an interpolation between the reconstructed hit position in ring B and the interaction vertex, separated for the up- and downstream side of the ring A counters in the kinematic range $0.08\,(\text{GeV/c})^2 < t < 0.64\,(\text{GeV/c})^2$. The quantity t denotes the square of the four-momentum transfer to the proton deduced with the kinematically constrained fit. The points of the upstream (downstream) side are artificially shifted by 2 cm to the right (left) for the purpose of visualisation. The horizontal error bars indicate the bin size in z

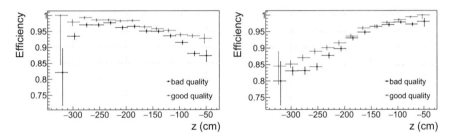

Fig. 5.17 The efficiency of ring A as a function of the z-position, given by an interpolation between the reconstructed hit position in ring B and the interaction vertex, for the upstream side (left) and the downstream side (right). The scintillators were separated according to high (good quality) or low (bad quality) attenuation lengths in the kinematic range $0.08\,(\text{GeV/c})^2 < t < 0.64\,(\text{GeV/c})^2$. The quantity t denotes the square of the four-momentum transfer to the proton, deduced with the kinematically constrained fit. The points of the bad (good) scintillators are artificially shifted by 2 cm to the right (left) for the purpose of visualisation. The horizontal error bars indicate the bin size in z

the attenuation length than others. These scintillators were marked of "bad quality" in the very beginning of the 2012 measurement. In order to account for this fact in the analysis, the efficiency is parametrised for each counter separately. The corresponding results can be seen in Appendix A.2.1 inside Figs. A.8 and A.13. A low attenuation length makes the adjustment of the high voltage in terms of the trade-off mentioned above very difficult. This fact was the major argument for the exchange of ring A for the 2016/2017 measurement, which is shortly covered in Chap. 9.

The dependence of the ring A efficiency on the beam charge is shown in Table 5.5. For the overall ring A efficiency a five percent effect is visible, which is taken into

Table 5.5 Ring A efficiencies for the two data yields of different beam charge integrated over the range $0.08\,(\text{GeV/c})^2 < t < 0.64\,(\text{GeV/c})^2$

	Upstream efficiency	Downstream efficiency	Ring A efficiency
μ^+ and μ^- yield	0.951 ± 0.001	0.927 ± 0.001	0.882 ± 0.002
μ^+ yield	0.935 ± 0.003	0.901 ± 0.003	0.842 ± 0.004
μ^- yield	0.957 ± 0.001	0.935 ± 0.002	0.895 ± 0.002

account in Chap. 7 as a correction factor for the two independent data yields of different beam charge. There is no reason why the beam charge itself could have an impact on the efficiency. However, the fact that the beam flux of the μ^+ beam was higher by about a factor of two compared to the μ^- beam might give an explanation. Regarding in addition the fact that Table 5.7 shows no such effect in case of ring B, the high occupancy in ring A, related to its close position to the beam, most likely causes a decrease of the efficiency as one increases the beam flux.

The efficiency of ring B as a function of t is shown in Table 5.6. The slight statistical tention with respect to the first bin in case of the upstream side determination is not further taken into account. The evolution of the ring B efficiency with the longitudinal hit position in Fig. 5.18 shows some slight trend for the downstream side and for this reason the z-dependence of the ring B efficiency is extracted for each counter of ring B separately, like it is done for ring A. The corresponding distributions are shown in Appendix A.2.2 inside Figs. A.19 and A.24. Furthermore, as mentioned above Table 5.7 shows no difference for the μ^+ and the μ^- data yield in case of ring B.

Table 5.6 Ring B efficiencies as a function of t, the square of the four-momentum transfer to the proton

$|t|$ range in $(\text{GeV/c})^2$	Upstream efficiency	Downstream efficiency	Ring B efficiency
]0.08, 0.22]	0.994 ± 0.001	0.990 ± 0.001	0.984 ± 0.002
]0.22, 0.36]	0.998 ± 0.001	0.992 ± 0.003	0.990 ± 0.002
]0.36, 0.5]	0.998 ± 0.001	0.994 ± 0.002	0.992 ± 0.003
]0.5, 0.64]	0.995 ± 0.003	0.990 ± 0.004	0.985 ± 0.004

Table 5.7 Ring B efficiencies for the two data yields of different beam charge integrated over the range $0.08\,(\text{GeV/c})^2 < t < 0.64\,(\text{GeV/c})^2$

	Upstream efficiency	Downstream efficiency	Ring B efficiency
μ^+ and μ^- yield	0.995 ± 0.001	0.991 ± 0.001	0.986 ± 0.002
μ^+ yield	0.993 ± 0.003	0.991 ± 0.003	0.984 ± 0.004
μ^- yield	0.996 ± 0.001	0.991 ± 0.002	0.988 ± 0.002

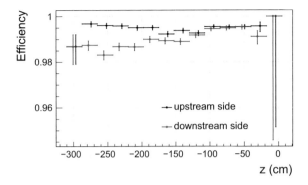

Fig. 5.18 The efficiency of ring B as a function of the z-position pinpointed by the interaction vertex and the proton momentum deduced by the kinematically constrained fit, separated for the up- and downstream side of the ring B counters in the kinematic range $0.08\,(\text{GeV/c})^2 < t < 0.64\,(\text{GeV/c})^2$. The quantity t denotes the square of the four-momentum transfer to the proton, deduced with the kinematically constrained fit. The points of the upstream (downstream) side are artificially shifted by 2 cm to the right (left) for the purpose of visualisation. The horizontal error bars indicate the bin size in z

References

1. K. Schmidt, Transverse target spin asymmetries in exclusive vector meson muoproduction, Dissertation, Albert Ludwigs Universität Freiburg (2014), https://freidok.uni-freiburg.de/data/ 9893, URN: urn:nbn:de:bsz:25-opus-98939
2. M. Gorzellik, Cross-section measurement of exclusive π^0 muoproduction and firmware design for an FPGA-based detector readout, Dissertation in preparation, Albert Ludwigs Universität Freiburg (2018)
3. R. Schäfer, Charakterisierung eines Detektors zum Nachweis von Rückstoßprotonen am COM-PASS Experiment, Diploma thesis, Albert Ludwigs Universität Freiburg (2013)
4. J. Pretz, Veto dead time and beam duty factor (2010). COMPASS Note 2010-8
5. NMC Collaboration, M. Arneodo et al., Measurement of the proton and deuteron structure functions, F_2^p and F_2^d, and of the ratio σ_L/σ_T. Nucl. Phys. B **483**, 3–43 (1997). https://doi. org/10.1016/S0550-3213(96)00538-X
6. SMC Collaboration, B. Adeva et al., Spin asymmetries A_1 and structure functions g_1 of the proton and the deuteron from polarized high-energy muon scattering. Phys. Rev. D **58**, 112001 (1998). https://doi.org/10.1103/PhysRevD.58.112001
7. E143 Collaboration, K. Abe et al., Measurements of $R = \sigma_L/\sigma_T$ for $0.03<x<0.1$ and fit to world data. Phys. Lett. B**452**, 194–200 (1999). https://doi.org/10.1016/S0370-2693(99)00244-0
8. S. Landgraf, Luminosity calculation for the COMPASS experiment using the F2 structure function, Master thesis, Albert Ludwigs Universität Freiburg (2015). Abbendum to luminosity calculation for the COMPASS experiment using the F2 structure function
9. E. Fuchey, Beam flux determination for the 2012 data using random triggers (2015). private communications
10. H. Wollny, Measuring azimuthal asymmetries in semi-inclusive deep-inelastic scattering off transversely polarized protons, Dissertation, Albert Ludwigs Universität Freiburg (2010), https://freidok.uni-freiburg.de/data/7558, URN: urn:nbn:de:bsz:25-opus-75589
11. O. Kouznetsov, Extraction of the badspill lists for the 2012 DVCS data (2015). Private communications

12. M. Büchele, Entwicklung eines FPGA-basierten 128-kanal time-to-digital converter für Teilchenphysik-experimente, Diploma thesis, Albert Ludwigs Universität Freiburg (2012)
13. E. Fuchey, Extraction of the bad spill list for the Fiber Station time jumps for the 2012 DVCS data (2015). Private communications

Chapter 6
Event Selection and Simulations

After a short introduction to the available simulation techniques the event selection of single exclusive photon production events is described. The chapter concludes with the application of the kinematically constrained fit to the single photon sample.

6.1 Overview of the Monte Carlo Simulations

Several Monte Carlo samples are available to describe the 2012 data. With respect to the nomenclature of Ref. [2, 10] the Monte Carlo productions, used in Sect. 6.2 and Chap. 7, are given by:

- LEPTO sample: Production_16-02 v1, LEPTO.
- Single photon sample: Production_16-02 v1, DVCS/BH.
- Exclusive π^0 sample: Production_16-02 v1, Pi0.

The production of a Monte Carlo sample at the COMPASS-II experiment can be split into three distinct steps: The event generation, the particle tracking through detector geometries and the treatment of Monte Carlo information and reconstruction.

6.1.1 Event Generation

The first step of the production of a Monte Carlo sample at the COMPASS-II experiment is the event generation. It involves the different types of event generators. The event generators produce the full set of kinematic variables for a given reaction, according to the underlying production mechanisms. Two different event generators are used throughout this thesis:

© Springer International Publishing AG, part of Springer Nature 2018
P. Jörg, *Exploring the Size of the Proton*, Springer Theses,
https://doi.org/10.1007/978-3-319-90290-6_6

LEPTO 6.1:

LEPTO 6.1 generates a variety of different particles produced in deep inelastic scattering processes and their corresponding kinematic properties. It accounts for semi-inclusive production mechanisms. The events are not weighted and the number of events in a certain region of the phase space is directly proportional to the cross sections of the individual processes. A complete description of the LEPTO generator can be found in Ref. [1]. The LEPTO Monte Carlo sample is used in this thesis to describe the semi-inclusive background contribution to the data. This background originates mainly from the production of π^0, decaying into two photons.

HEPGen++:

This generator has been specifically developed for the COMPASS-II experiment in order to account for different exclusive production mechanisms. A detailed description of the event generator can be found in Ref. [2]. It is a weighted event generator. Though the number of produced events inside a certain phase space element is kept close to the expected number of real data events, it is the event weight which accounts for the shape of the underlying cross section. Throughout this analysis HEPGen++ is used for the generation of an exclusive single photon and an exclusive π^0 Monte Carlo sample.

For the pure Bethe-Heitler cross section there is a precise calculation available, which includes the mass of the muon in the propagator [3]. The formula was developed by P. A. M. Guichon and cross checked with an analytic and a numeric approach. The corresponding event weight was introduced into HEPGen++ and shall be referred to as $w_{P.A.M.}$ in the following.

In case of the full exclusive single photon cross section, including the DVCS process and the interference term, three weighting factors are available. Each generated event is assigned a weight, which accounts for the DVCS (w_{DVCS}) and the Bethe-Heitler process (w_{BH}) as well as the interference term (w_I). The final weight w is then given by:

$$w = w_{BH} + w_{DVCS} + w_I. \tag{6.1}$$

The calculation of these weighting factors goes back to the DVCS model of Frankfurt, Freund and Strikman [4, 5]. It has been adapted by A. Sandacz [6] to introduce the Bethe-Heitler calculations from Ref. [7], while the propagators were recalculated by P. A. M. Guichon, to include the lepton/muon mass. The t-dependence B of the DVCS cross section has been parametrised in the following way:

$$B(x_{Bj}) = B_0 + 2\alpha' \ln\left(\frac{x_0}{x_{Bj}}\right),$$

The parameters (B_0, α', x_0) describe the x_{Bj}-dependence of B. They have been chosen as follows:

$$(B_0, \alpha', x_0) = \left(4.942\,(\text{GeV}/c)^{-2}, 0.8\,(\text{GeV}/c)^{-2}, 0.042\right).$$

It is the only available calculation, which accounts for the full cross section of exclusive single photon production, including the mass of the muon at least in an approximate way.

The basic difficulty within a calculation of the full exclusive single photon cross section occurs within the Bethe-Heitler cross section. The Bethe-Heitler cross section changes very rapidly in case the photon is emitted along the direction of the incident or in the opposite direction of the scattered muon, which is known as s and p peak in the literature [8]. The underlying problematic is related to helicity conservation. It can be illustrated by imagining a massless lepton with helicity $\frac{1}{2}$ in the initial state, which radiates a real photon in the very same direction it is travelling. Helicity conservation would force the photon to have a helicity of zero. This is forbidden for a real photon and would cause a singularity in the cross section. Especially when it comes to the calculation of the interference term between the DVCS and the Bethe-Heitler process, it is a non trivial issue to introduce the muon mass into the different calculations of the exclusive single photon cross section available on the market. These calculations have neglected the lepton mass so far, due to the fact that all DVCS experiments apart from COMPASS-II are making use of an electron beam.

The event by event weights for the exclusive π^0 Monte Carlo sample are generated with HEPGen++ according to the model of Goloskokov and Kroll. A detailed description of the model and the used GPD parameters are given in Refs. [2, 9]. The corresponding weighting factors shall be denoted as w_{π^0} in the following.

6.1.2 Particle Tracking Through Detector Geometries

During the step of particle tracking through Detector Geometries the precise description of the detector geometries and material composition is introduced into the simulations. Basically, vertices inside the target are created, while the kinematic properties of the beam particle are taken from a so called beam file and passed to the event generator. The event generator creates the outgoing particles which traverse a complete Geant4[1] simulation of the COMPASS-II spectrometer. This simulation accounts for effects such as e^+e^- pair production, energy loss of charged particles in the materials, bending of charged tracks in a magnetic field and hadronic interactions, just to mention a few. The hit positions and depending on the detector type the energy deposit of particles within the variety of detectors of the COMPASS-II spectrometer is collected. While a complete description of the TGEANT[2] software is given within Ref. [10], a few aspects dedicated to the 2012 data taking shall be highlighted at this stage:

- Each scintillator of the CAMERA detector is aligned within the simulations according to the calibration constants extracted from the 2012 data.

---

[1]**GE**ometry **AN**d **T**racking.
[2]**T**otal **GE**ometry **AN**d **T**racking.

- As it will be shown in Sect. 6.2.1 the liquid hydrogen target was slightly declined along the beam axis, which is also taken into account in the simulations.
- The extraction of a beam file from real data, which describes the full phase space of the μ^+ and respectively μ^- beam, has been performed.
- The pile up and halo contribution was extracted from real data and introduced into the simulations according to the measured beam flux, which differs for the μ^+ and μ^- data taking periods.

6.1.3 Treatment of Monte Carlo Information and Reconstruction

The final step of a Monte Carlo production can be separated into the treatment of Monte Carlo information and the reconstruction.

After the particle tracking stage in Monte Carlo is completed, the intersection points and the energy deposit in the various detectors are digitised to form so called hits. Furthermore, uncertainties and efficiency corrections [2] are added in order to account for the different detection mechanisms of the various detector types. In certain cases information can also be added at this stage, as it is the case for the introduction of a noise contribution into the collected calorimeter data [2]. In the COMPASS-II collaboration these steps are not performed at Monte Carlo generation level, but later during the reconstruction of Monte Carlo data. This provides the advantage that one can easily change the uncertainties and efficiency corrections without a time consuming reproduction of the Monte Carlo information. The final reconstruction step is completely analogue for real and Monte Carlo data. It is shortly covered in Sect. 2.7.

In the following the treatment of the Monte Carlo information in case of CAMERA shall be demonstrated. Because the scintillators of the detector were placed according to the calibration constants extracted from the 2012 data, the treatment is straight forward. It shall be explained for an exemplary ring B scintillator i. The particle tracking stage provides the absolute longitudinal hit position z_{Bi} of particles traversing the scintillator together with the absolute time T_{Bi} at which the particle has crossed the element. For the real data reconstruction the longitudinal hit position z_{Bi} is constructed from the time-stamps $t_{Bi}^{u,d}$ of a photomultiplier signal at the upstream (u) and downstream (d) side of the scintillator as follows:

$$z_{Bi} = \frac{1}{2}c_{Bi}(t_{Bi}^u - t_{Bi}^d) + k_{Bi}^z. \tag{6.2}$$

The effective speed of light c_{Bi} and the calibration constant k_{Bi}^z are given according to Sect. 5.1.4. Solving Eq. (6.2) together with:

$$T_{Bi} = \frac{(t_{Bi}^u + t_{Bi}^d)}{2},$$

for the time-stamps $t_{Bi}^{u,d}$ yields:

$$t_{Bi}^{u} = \frac{z_{Bi} - k_{Bi}^{z}}{c_{Bi}} + T_{Bi} \text{ and } t_{Bi}^{d} = -\frac{z_{Bi} - k_{Bi}^{z}}{c_{Bi}} + T_{Bi}. \tag{6.3}$$

The uncertainties on the time-stamps $\sigma(t_{Bi}^{u,d})$ can be calculated from Eq. (6.2) as follows:

$$\sigma(t_{Bi}^{u,d}) = \frac{\sqrt{2}}{c_{Bi}} \sigma_{z}(B), \tag{6.4}$$

For this the assumption $\sigma(t_{Bi}^{u}) = \sigma(t_{Bi}^{d}) := \sigma(t_{Bi}^{u,d})$ has been made. The same relations hold for the inner ring of scintillators by replacing B_i with A_i in Eqs. (6.2)–(6.4). The quantities $\sigma_z(A)$ and $\sigma_z(B)$ coincide with the values given in Sect. 4.5.

The time-stamps $t_{Bi}^{u,d}$ and $t_{Ai}^{u,d}$ of Eq. (6.3) are smeared randomly for each Monte Carlo hit according to a Gaussian distribution. The width of this distribution is given by Eq. (6.4). The fact that certain segments of the CAMERA detector had to be disabled during the data taking, is taken into account in the simulations by disabling a segment of the detector with the probability given by Eq. (5.11). Furthermore, the efficiency of each of the 48 scintillators is introduced individually as a function of the longitudinal hit position, according to Sect. 5.4.

The resulting time-stamps can be treated in the same way as the corresponding time-stamps for real data. The only exception is that the time of flight calibration constants of Sect. 5.1.5 are put to zero by default in case of the reconstruction of Monte Carlo data.

6.2 Event Selection of Exclusive Single Photons

The event selection is supposed to be sensitive on exclusive single photon production, $\mu p \rightarrow \mu' \gamma p'$, without excluding events associated with background due to pile up. A summarised presentation of the event selection is given in Table 6.1.

To guarantee a stable beam, events occurring within 1 and 10.4 s with respect to the begin of a spill are considered. Furthermore, only those events, which have been triggered by the Middle, Ladder or Outer trigger are considered within the analysis.[3] The further event selection can be split into the following three main steps:

- **Muon and vertex selection**:
  For each event all primary vertices, which satisfy the criteria of Sect. 6.2.1, are considered. A primary vertex denotes a vertex which includes the beam particle
- **Photon selection**:
  If the event contains a single neutral cluster, which satisfies the criteria of Sect. 6.2.2, the event is further considered.

---

[3]These are all the relevant physics triggers being active for the 2012 data taking apart from the LAS trigger, which was strongly prescaled.

Table 6.1 Overview of the selection of exclusive single photon events

Events with:	General event criteria
<ul><li>Time in spill: $1\,\text{s} < T < 10.4\,\text{s}$</li><li>Considered trigger types: Middle Trigger (MT) or Ladder Trigger (LT) or Outer Trigger (OT)</li></ul>	
Primary vertices with:<ul><li>Vertex z-position: $-311.2\,\text{cm} < v_z < -71.2\,\text{cm}$</li><li>Vertex distance from target centre: $d < 1.9\,\text{cm}$ *(see Sect. 6.2.1)*</li><li>One incoming charged track μ with: >2 hits in the Beam Momentum Stations (BMS), >1 hit in the Scintillating Fibre detectors (Fi), >2 hits in the Silicon detectors (Si), beam momentum: $140\,\text{GeV/c} < p_\mu < 180\,\text{GeV/c}$, beam track traverses the full target volume *(see Sect. 6.2.1)*</li><li>One outgoing charged track μ' with: same charge than incoming track, traversed radiation lengths: $X/X_0 > 15$, z-position of first measured point: $z_{first} < 350\,\text{cm}$, z-position of last measured point: $z_{last} > 350\,\text{cm}$,</li><li>Inclusive scattering variables: energy loss: $10\,\text{GeV} < \nu < 32\,\text{GeV}$, photon virtuality: $1\,(\text{GeV/c})^2 < Q^2 < 5\,(\text{GeV/c})^2$</li></ul>	(1) Muon and vertex selections
Exactly one neutral cluster γ with:<ul><li>A valid cluster time *(see Sect. 6.2.2)*</li><li>A reconstructed cluster energy: $E_\gamma > 4,\ 5,\ 10\,\text{GeV}$ in ECal0, 1, 2</li></ul>	(2) Photon selections
Reconstructed CAMERA tracks with:<ul><li>Longitudinal hit position z inside ring A and B: $-366.19\,\text{cm} < z_A < 8.81\,\text{cm}$, $-338.94\,\text{cm} < z_B < 71.06\,\text{cm}$</li><li>Velocity of reconstructed recoiling particle: $0.1 < \beta := \frac{v}{c} < 1$</li></ul>	(3) CAMERA selections
All combinations of (1), (2) and (3) which satisfy:<ul><li>$\lvert \Delta p_T \rvert < 0.3\,\text{GeV/c}$</li><li>$\lvert \Delta \phi \rvert < 0.4\,\text{rad}$</li><li>$\lvert \Delta z_A \rvert < 16\,\text{cm}$</li><li>$\lvert M_X^2 \rvert < 0.3\,\left(\text{GeV/c}^2\right)^2$ *(see Sect. 6.2.3)*</li><li>Exactly one combination must be left</li><li>Square of the proton four-momentum transfer: $0.08\,(\text{GeV/c})^2 < t < 0.64\,(\text{GeV/c})^2$</li><li>Remove visible leaking π^0 contribution: $m_{\gamma\gamma} < 115\,(\text{MeV/c}^2)$ or $m_{\gamma\gamma} > 155\,(\text{MeV/c}^2)$ *(see Sect. 7.2)*</li></ul>	Exclusivity selections

- **Proton selection and application of the exclusivity cuts**:
  In case there is at least one vertex, satisfying the muon and vertex selection criteria, and exactly one neutral cluster, satisfying the photon selection criteria, all combinations with the reconstructed tracks inside CAMERA are considered. For each of those combinations the exclusivity variables described in Sect. 6.2.3 are calculated. If a single combination remains after cutting on the exclusivity variables, the event is considered to be an exclusive single photon event.

In the following plots data from the 2012 run are shown and compared to Monte Carlo samples. The distributions denoted with "π^0 background" are estimated according to Sect. 7.2. If not denoted otherwise, the overall Monte Carlo prediction, indicated by the red shaded distributions, is the sum of the π^0 background estimate and the single photon Monte Carlo yield.

The single photon Monte Carlo yield is normalised to the luminosity of the 2012 data. To achive this, the Monte Carlo luminosity $\mathcal{L}_{MC}$ has to be calculated according to:

$$\mathcal{L}_{MC} = \frac{\sum_{\Delta\Omega} w_{\mathrm{DVCS}}}{\int_{\Delta\Omega} \left(\frac{d\sigma_{\mathrm{HEPGen++}}^{\mathrm{DVCS}}}{d\Omega} \right) d\Omega} \tag{6.5}$$

Here, $\left(\frac{d\sigma_{\mathrm{HEPGen++}}^{\mathrm{DVCS}}}{d\Omega} \right)$ denotes the differential DVCS cross section as it is included in HEPGen++. The phase space region $\Delta\Omega$ is given by $(\Delta Q^2 \Delta\nu \Delta t \Delta\phi_{\gamma^*\gamma})$. It can be choosen arbitrarily. However, the crucial point is that the sum over the generated event weights in the numerator of Eq. (6.5) covers exactly the same phase space region $\Delta\Omega$ as the integration in the denominator. In principle one can replace "DVCS" by "BH" or "P.A.M." in Eq. (6.5). The outcome would be the same, because the weights in the numerator are calculated according to the differential cross section of the denominator. Thus, in order to normalise Monte Carlo to data the Monte Carlo has to be scaled by:

$$N_{MC} = \frac{\mathcal{L}}{\mathcal{L}_{MC}}, \tag{6.6}$$

while $\mathcal{L}$ denotes the luminosity of the data. The calculation of $\mathcal{L}$ is outlined in Sect. 5.2.

After this normalisation procedure the following event weight is used:

$$w = w_{\mathrm{BH}} + 0.6\, w_{\mathrm{DVCS}} + \sqrt{0.6}\, w_{\mathrm{I}}. \tag{6.7}$$

Equation (6.7) is modified by a fudge factor of 0.6 with respect to Eq. (6.1). This accounts for the fact that the DVCS contribution is overestimated by the used DVCS model. It should be emphasised that this rescaling of the DVCS model is not used for the extraction of any "physics" quantity. It has the simple purpose to get a better visual agreement between data and Monte Carlo, by using one single normalisation procedure. Furthermore, all following distributions show the resulting quantities corrected by the kinematically constrained fit, which will be described in Sect. 6.3.

The only exception is given by the exclusivity distributions of Sect. 6.2.3, which would simply be zero after the application of the fitting procedure.

6.2.1 Muon and Vertex Selection

All interaction vertices of the beam particle within the liquid hydrogen target, which provide a single in- and outgoing track of equal charge, are considered.

The precise location of the target cell has been extracted using a dedicated event selection of vertices with more than two outgoing particles. With this requirement it was possible to extract the position of the target window mylar directly from the data. The procedure is described in detail in Ref. [11] and the final parametrisation shown in the right graph of Fig. 6.1 was provided by Ref. [12]. It shows the extracted x and y-position of the target cell centre together with the extracted target cell radius r as a function of the longitudinal z-coordinate.

In the analysis all vertices within a radius of 1.9 cm with respect to the target cell centre are considered. The left distribution of Fig. 6.1 shows the distribution of the longitudinal vertex position v_z of the final event sample. Vertices which satisfy the condition indicated by the blue dotted lines:

$$-311.2\,\text{cm} < v_z < -71.2\,\text{cm},$$

are considered.

For a correct determination of the luminosity it is also required that the extrapolation of the incoming beam track crosses the full target volume. Figure 6.2 shows

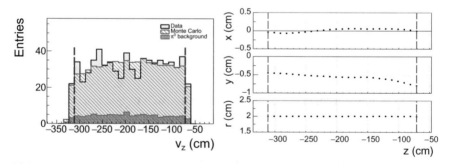

Fig. 6.1 Left: Distribution of the longitudinal vertex position v_z of the final event sample, used to extract the DVCS cross section. Vertices which satisfy the condition indicated by the blue dotted lines are considered. Right: Parametrisation of the target cell extracted from the data and provided by Ref. [12]. The extracted x- and y-position of the target cell center together with the extracted target cell radius r as a function of the longitudinal z-coordinate are shown. The left distribution is shown after the full event selection, disabling the cut on v_z. The corresponding distributions for an extended kinematic range are shown in Fig. A.26

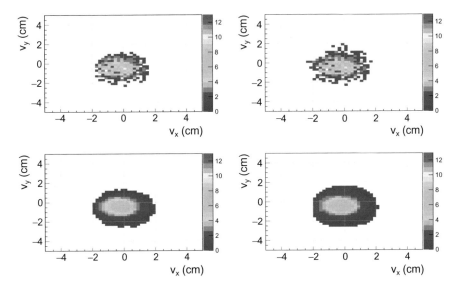

Fig. 6.2 The x and y-position of the interaction vertex $\vec{v}$ before (right) and after (left) the application of the target crossing requirement for data (top) and Monte Carlo (bottom). All distributions are shown after the full event selection disabling the cut on the respective variable if applicable. The corresponding distributions for an extended kinematic range are shown in Fig. A.26

the x- and y-position of the interaction vertex $\vec{v}$ before and after the application of the target crossing requirement for data and Monte Carlo.

In order to provide a precise measurement of the momentum vector of the beam particle by the so called "Beam Telescope", it is ensured that at least three hits in the Beam Momentum Stations (BMS), at least three hits in the Silicon detectors, and at least two hits in the Scintillating Fibre detectors upstream of the target have been measured. Furthermore, the momentum of the incoming beam particle, p_μ has to satisfy the condition:

$$140\,\text{GeV/c} < p_\mu < 180\,\text{GeV/c},$$

indicated by the blue lines inside the top left distribution of Fig. 6.3.

The outgoing charged particle is required to have traversed more than 15 radiation lengths to be identified as a muon. For a precise determination of the momentum of the scattered muon at least one hit is required on either side upstream and downstream of the first spectrometer dipole. The momentum of the outgoing particle $p_{\mu'}$, its polar angle $\theta_{\mu'}$ and its azimuthal angle $\phi_{\mu'}$ in the laboratory frame are shown in Fig. 6.3. The hole at $\phi_{\mu'} \approx \pm\pi$ and the decrease at $\phi_{\mu'} = 0$ in case of the bottom left distribution of Fig. 6.3 are related to the kinematic coverage of the trigger hodoscopes. The positioning of the trigger hodoscopes is described in Sect. 3.6.1.

The Lorentz invariant quantity ν, which coincides with the energy loss of the muon in the laboratory system, is shown in Fig. 6.4. In order to select a phase space

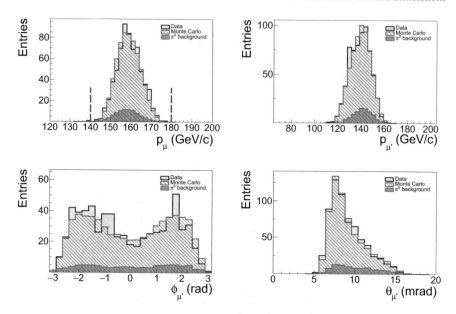

Fig. 6.3 Distributions of the in- and outgoing muon. Top left: Distribution of the momentum of the incoming muon p_μ. Top right: Distribution of the momentum of the outgoing muon $p_{\mu'}$. Bottom left: Distribution of the polar angle $\theta_{\mu'}$ of the momentum vector of the outgoing muon. Bottom right: Distribution of the azimuthal angle $\phi_{\mu'}$ of the momentum vector of the outgoing muon. All distributions are shown after the full event selection disabling the cut on the respective variable if applicable. The corresponding distributions for an extended kinematic range are shown in Fig. A.27

region for which the DVCS process becomes sizeable, small values of ν have to be considered. This will be demonstrated in detail in Sect. 7.3. The selected ν-region, for which the DVCS cross section is extracted, is indicated by the blue lines of Fig. 6.4. The blue lines satisfy the condition:

$$10\,\text{GeV} < \nu < 32\,\text{GeV}.$$

The distributions of Q^2 and the Bjorken scaling variable x_{Bj} are shown in Fig. 6.5. The selected region of Q^2 is illustrated by the blue lines and satisfies the condition:

$$1\,(\text{GeV/c})^2 < Q^2 < 5\,(\text{GeV/c})^2.$$

The lower boundary for Q^2 is motivated by "physics", in order to apply the factorisation theorem, mentioned in Sect. 2.3.1. For the upper boundary condition, it is in principle desirable to enlarge the analysis range to larger Q^2. In Sect. 7.4 it is shown that this is unfortunately not possible for the 2012 data.

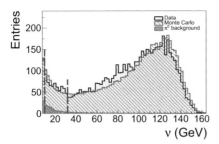 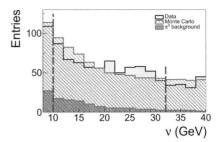

Fig. 6.4 Distributions of the Lorentz invariant quantity ν, which coincides with the energy loss of the muon in the laboratory system. Left: Distribution for the full single photon sample. Right: A zoom on the region indicated by the blue lines inside the left distribution. In order to select a phase space region for which the DVCS process becomes sizeable, small values of ν have to be considered, which are indicated by the blue lines. All distributions are shown after the full event selection disabling the cut on the respective variable if applicable. The corresponding distributions for an extended kinematic range are shown in Fig. A.28

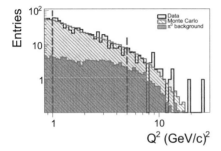 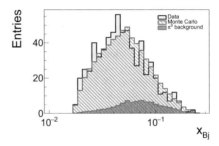

Fig. 6.5 Distributions of the photon virtuality Q^2 (left) and the Bjorken scaling variable x_{Bj} (right). The applied cut is indicated by the blue lines. All distributions are shown after the full event selection disabling the cut on the respective variable if applicable. The corresponding distributions for an extended kinematic range are shown in Fig. A.28

6.2.2 Photon Selection

All clusters measured by the three electromagnetic calorimeters which are not associated to a charged track are considered to be photons. For the single photon selection the reconstructed cluster energy E_γ has to satisfy the relation:

$$E_\gamma > 4,\ 5,\ 10\,\text{GeV in ECal0, ECal1, ECal2,}$$

while "ECal n" denotes one of the three electromagnetic calorimeters of the COMPASS-II spectrometer. The values of the thresholds have been evaluated using the distributions shown in Fig. 6.6. These distributions have been derived by applying the event selection, without the application of a photon threshold, to a single

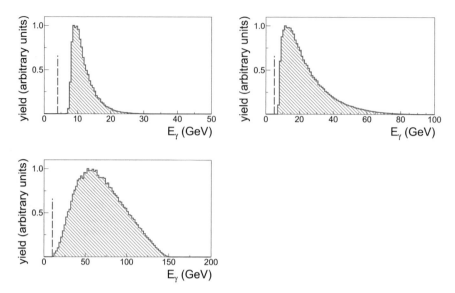

Fig. 6.6 Distributions showing the reconstructed photon energy in ECal0 (top left), ECal1 (top right) and ECal2 (bottom left), derived by applying the event selection to a single photon Monte Carlo yield, while no threshold for the photon was applied. The event weight is given by w_{DVCS}, which corresponds to the hypothesis of a pure DVCS cross section. The photon energy threshold is shown by the blue lines. The enlarged kinematic range for this study is: $0.08\,(\text{GeV/c})^2 < |t| < 0.64\,(\text{GeV/c})^2$, $8\,\text{GeV} < \nu < 32\,\text{GeV}$, $1\,(\text{GeV/c})^2 < Q^2 < 20\,(\text{GeV/c})^2$

photon Monte Carlo yield. In this case the event weight was chosen to accord to the hypothesis of a pure DVCS cross section.

The time of the neutral cluster with respect to the trigger time is examined as a function of the cluster energy. In case the cluster timing is outside the blue bands, shown in Fig. 6.7, the cluster is rejected. The blue bands have been extracted using a dedicated event selection, which provides a large amount of reconstructed calorimeter clusters. The parametrisations correspond to three sigma bands and have been provided by Ref. [13].

The distributions of the magnitude and the polar and azimuthal angle of the photon momentum in the laboratory frame are shown in Fig. 6.8. The hole at $\phi_\gamma = 0$ and the decrease at $\phi_\gamma \approx \pm\pi$ in the top right distribution of Fig. 6.8 are directly related to the corresponding distribution of the scattered muon. For exclusive single photon production most of the outgoing momentum is carried by the scatterd muon and the photon. Apart from the small contribution of the recoiled proton, the photon travels in the opposite hemisphere of the scattered muon. Thus, shifting the distribution of the scattered muon (bottom left side of Fig. 6.3) by π results approximately in the top right side of Fig. 6.8.

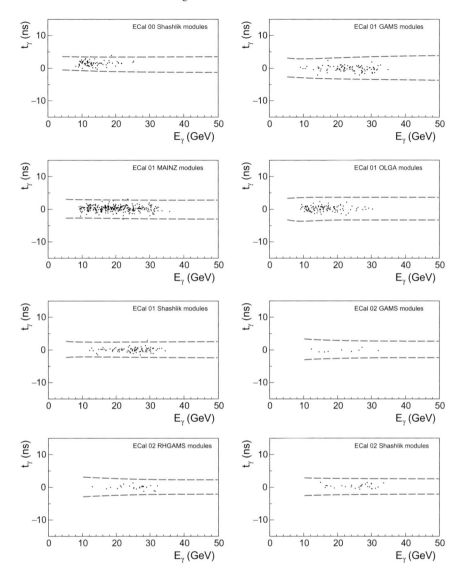

Fig. 6.7 Two dimensional distributions of the time of the reconstructed neutral calorimeter cluster with respect to the trigger time as a function of the reconstructed cluster energy for the different calorimeter cell types of the three electromagnetic calorimeters. The different cell types and calorimeters are indicated within the distributions. The blue three sigma bands indicate the applied cuts. All distributions are shown after the full event selection disabling the cut on the calorimeter timing. The corresponding distributions for an extended kinematic range are shown in Fig. A.29

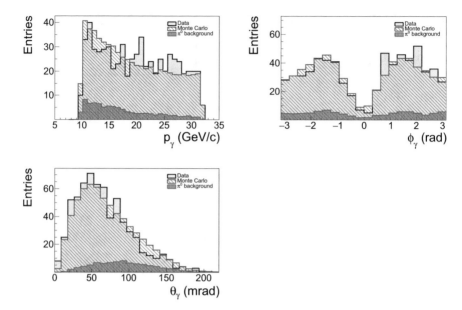

Fig. 6.8 Distributions of the magnitude p_γ (top left), the azimuthal ϕ_γ (top right) and polar θ_γ (bottom left) angle of the photon momentum after the application of the full event selection. The corresponding distributions for an extended kinematic range are shown in Fig. A.30

6.2.3 Proton Selection and Application of the Exclusivity Cuts

All reconstructed tracks inside the CAMERA detector which provide a longitudinal hit position $z_{A;B}$ in ring A and ring B of:

$$-366.19\,\text{cm} < z_A < 8.81\,\text{cm},$$

$$-338.94\,\text{cm} < z_B < 71.06\,\text{cm},$$

are considered. In addition, the reconstructed velocity in units of the speed of light associated to these tracks has to satisfy:

$$0.1 < \beta := \frac{v}{c} < 1.$$

The remaining tracks are combined with all vertices, passing Sect. 6.2.1 and the single photon of Sect. 6.2.2. Denoting the four-momenta of the beam and scattered muon as $p_\mu = (E_\mu/c, \vec{p}_\mu)$ and $p_{\mu'} = (E_{\mu'}/c, \vec{p}_{\mu'})$, the four-momenta of the initial and final state proton as $p_p = (m_p c, \vec{0})$ and $p_{p'} = (E_{p'}/c, \vec{p}_{p'})$ and the four-momentum of the photon as $p_\gamma = (E_\gamma/c, \vec{p}_\gamma)$, the hypothesis of exclusivity for the reaction $\mu p \to \mu' p' \gamma$ is tested with the following exclusivity variables.

- **Reverse vertex pointing**:
  The reconstructed interaction vertex together with the longitudinal hit position in the outer ring of the CAMERA detector allows for an interpolation, which yields the longitudinal hit position $z_{A,interp.}$ in the inner ring. The interpolated hit position is compared to $z_{A,reco}$, the reconstructed hit position in the inner ring. This yields the quantity:

$$\Delta z_A = z_{A,interp.} - z_{A,reco}, \tag{6.8}$$

and the following cut is performed:

$$|\Delta z_A| < 16\,\text{cm}.$$

The procedure is analog to the calibration of the longitudinal position of the ring A counters. It is schematically illustrated in Fig. 5.7.

- **Missing mass**: The detection of the proton in the CAMERA detector allows performing a cut on the square of the missing mass M_X^2 of an additional particle. This corresponds essentially to a check of the exclusivity by exploiting the four-momentum balance of the reaction:

$$M_X^2 c^2 = (p_{\mu'} + p_p - p_{\mu'} - p_{p'} - p_\gamma)^2 = 2(m_p c^2 - E_{p'})(\nu - E_\gamma - E_{p'}) + tc^2, \tag{6.9}$$

and the following cut is performed:

$$|M_X^2| < 0.3 \left(\text{GeV/c}^2\right)^2.$$

It is worth to emphasise that the quantities $t = (p_p - p_{p'})^2$ and $E_{p'}$ are calculated from the reconstructed proton momentum inside CAMERA by assuming the mass of the proton.

- **Coplanarity**:
  Using the beam and spectrometer measurements, the momentum of the recoiled particle can be predicted as:

$$\vec{p}_{pred} = \vec{p}_\mu - \vec{p}_{\mu'} - \vec{p}_\gamma. \tag{6.10}$$

This yields the predicted azimuthal angle of the momentum of the recoiled particle $\phi_{pred.}$. It is compared to the reconstructed azimuthal angle $\phi_{reco.}$ within the CAMERA detector. Thus, the following exclusivity variable allows performing a test on the coplanarity of the exclusivity hypothesis:

$$\Delta\phi = \phi_{pred.} - \phi_{reco.}, \tag{6.11}$$

and the following cut is performed:

$$|\Delta\phi| < 0.4\,\text{rad}.$$

- **Transverse momentum balance**:
  The transverse component $(\vec{p}_{pred})_T$ according to Eq. (6.10) is used. It is compared to $(\vec{p}_{reco.})_T$, the transverse component of the reconstructed momentum of the recoiled particle within the CAMERA detector. The result is the following exclusivity variable:

$$\Delta p_T = (\vec{p}_{pred})_T - (\vec{p}_{reco.})_T, \tag{6.12}$$

and the following cut is performed:

$$|\Delta p_T| < 0.3\,\text{GeV/c}.$$

The distributions of the four exclusivity variables are shown in Fig. 6.9. After the exclusivity cuts have been applied, it may happen in rare cases that a single event possesses still more than one combination of a vertex, a CAMERA track and the single photon. These ambiguous events are rejected. The number of ambiguous events with respect to the final event yields are between one and two percent with no noticeable difference for the beam charge. As one applies the same cuts to a single photon Monte Carlo sample, the number of ambigious events is also approximately two percent.

The square of the four-momentum transfer to the proton t and the polar angle of the proton momentum vector θ_p are shown in Fig. 6.10. It should be emphasised that the exact evolution as a function of $|t|$, corrected for the amount of the Bethe-Heitler contribution and the π^0 contamination, is unknown and subject to the next chapter. The comparison of data and Monte Carlo is shown at this stage to demonstrate a sufficient agreement in order to compute acceptance correction factors as a function of $|t|$. The calculation of the acceptance will be demonstrated in Sect. 7.4.

The measurement of the azimuthal angle of the recoiled proton with the CAMERA detector is achieved by 48 scintillating counters. Hence, it is more meaningfull to show Fig. 6.11. It illustrates the number of events in each of the 24 scintillating counters of ring A and ring B separately. As demonstrated in Sect. 5.4, the efficiency of CAMERA is extracted for each scintillator individually. Furthermore, as shown in Sect. 5.3, certain scintillators had to be excluded for certain runs, due to bit-flips on the ADCs of the readout electronic. Both effects were included in the simulations and explain the large fluctuations for the data and the Monte Carlo yield. Furthermore, in order to overcome statistical fluctuations the full single photon yield, including large values of ν, is shown in Fig. 6.11.

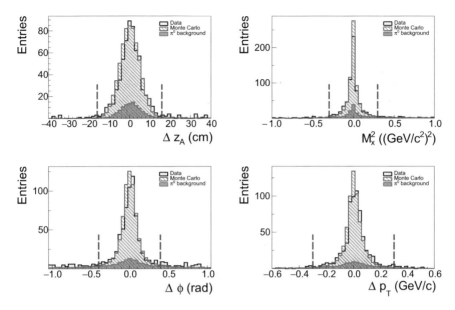

Fig. 6.9 Distributions of the exclusivity variables defined in Eqs. (6.8)–(6.12). The applied cuts are indicated by the blue lines. All distributions are shown after the full event selection disabling the cut on the respective variable if applicable. The corresponding distributions for an extended kinematic range are shown in Fig. A.31

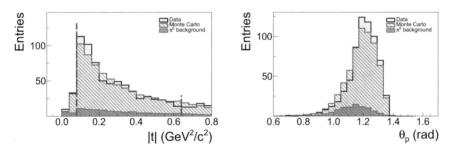

Fig. 6.10 Left: Distribution of the square of the four-momentum transfer to the proton t. Right: Distribution of the polar angle of the momentum vector of the recoiled proton with respect to the spectrometer coordinate system. All distributions are shown after the full event selection disabling the cut on the respective variable if applicable. The corresponding distributions for an extended kinematic range are shown in Fig. A.32

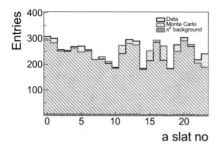
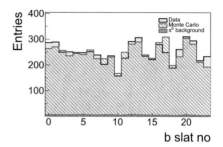

Fig. 6.11 Number of events detected in each of the 24 scintillating counters of ring A (left) and ring B (right). An extended kinematic range is shown: $10\,\text{GeV} < \nu < 144\,\text{GeV}$, $1\,(\text{GeV/c})^2 < Q^2 < 20\,(\text{GeV/c})^2$. The distribution in the kinematic range, used for the extraction of the DVCS cross section is shown in Fig. A.33

6.3   The Kinematic Fit for DVCS

With increasing values of $|t|$ the resolution of CAMERA gets worse quiet rapidly, while the resolution of the spectrometer improves. That is why there is a particular interest in a kinematic fit to provide a consistent solution to the most precise determination of $|t|$, measured by CAMERA together with the spectrometer.

The measured beam, spectrometer and CAMERA quantities for the kinematic fit are:

$$
\vec{k} = \begin{pmatrix} k_1 \\ \cdot \\ \cdot \\ \cdot \\ k_{23} \end{pmatrix} := \begin{pmatrix} \vec{p}_p \\ \vec{0}_{20} \end{pmatrix} + \begin{pmatrix} \vec{0}_3 \\ \vec{a}_\gamma \\ |\vec{p}_\gamma| \\ \vec{0}_{14} \end{pmatrix} + \begin{pmatrix} \vec{0}_6 \\ \vec{a}_\mu \\ \vec{p}_\mu \\ \vec{a}_{\mu'} \\ \vec{p}_{\mu'} \\ \vec{0}_7 \end{pmatrix} + \begin{pmatrix} \vec{0}_{16} \\ r_A \\ \phi_A \\ z_A \\ r_B \\ \phi_B \\ z_B \\ |\vec{p}_{p'}| \end{pmatrix}.
$$

The unmeasured quantities are:

$$
\vec{h} = \begin{pmatrix} h_1 \\ \cdot \\ \cdot \\ \cdot \\ h_7 \end{pmatrix} := \begin{pmatrix} \vec{v} \\ \vec{0}_4 \end{pmatrix} + \begin{pmatrix} \vec{0}_3 \\ \Theta_\gamma \\ \phi_\gamma \\ \vec{0}_2 \end{pmatrix} + \begin{pmatrix} \vec{0}_5 \\ \Theta_{p'} \\ \phi_{p'} \end{pmatrix}. \tag{6.13}
$$

The used abbreviations are explained in the following:

- The neutral element of $\mathbb{R}^N$ is depicted by $\vec{0}_N$.
- The target proton is assumed to be at rest and its momentum is denoted by $\vec{p}_p = \vec{0}$.

- The transverse coordinates and the magnitude of the momentum of the photon are given by $(\vec{a}_\gamma, |\vec{p}_\gamma|)^T$, which are treated together with the unmeasured photon parameters $(\Theta_\gamma, \phi_\gamma)^T$ according to Sect. 3.4.
- The quantities $(\vec{a}_\mu, \vec{p}_\mu)^T$ and $(\vec{a}_{\mu'}, \vec{p}_{\mu'})^T$ denote the track parameters of the beam particle and the scatterd muon. They are treated as described in Sect. 3.3.
- The parameters $(r_A, \phi_A, z_A, r_B, \phi_B, z_B, |\vec{p}_{p'}|)^T$ together with the unmeasured quantities $(\Theta_{p'}, \phi_{p'})^T$ describe the final state proton as explained in Sect. 3.5.
- The vertex position is depicted by $\vec{v}$.

The kinematic fitter then calculates corrections $\Delta \vec{k}$ to the measured quantities $\vec{k}$ such that the corrected measurements:

$$\vec{k}^{fit} = \vec{k} + \Delta \vec{k},$$

together with the unmeasured quantities $\vec{h}$ minimise the least squares function of Eq. (4.1). The minimisation is performed with subject to the constraints listed in the following:

The energy and momentum conservation constraints are given, according to Sect. 4.6.1, by:

$$
\begin{aligned}
g_i &= (p_\mu^{fit})_i - (p_{\mu'}^{fit})_i - (p_\gamma^{fit})_i - (p_{p'}^{fit})_i = 0, \\
g_4 &= E_\mu^{fit} + m_p c^2 - E_{\mu'}^{fit} - E_\gamma^{fit} - E_{p'}^{fit} = 0,
\end{aligned}
\tag{6.14}
$$

$\forall\, i \in \{1, 2, 3\}$, while the index denotes Cartesian components of the three-vectors.

The variables denoted with the superscript "fit" emphasise the fact that the quantities corrected by the kinematic fit have to satisfy the constraints. Apart from the energy and momentum conservation all tracks except the initial and final state proton must originate from a common vertex:

$$g_{4+i} = (p_\mu^{fit})_3 \left(v_i - (a_\mu^{fit})_i \right) - (p_\mu^{fit})_i \left(v_3 - (a_\mu^{fit})_3 \right) = 0,$$

$$g_{6+i} = (p_{\mu'}^{fit})_3 \left(v_i - (a_{\mu'}^{fit})_i \right) - (p_{\mu'}^{fit})_i \left(v_3 - (a_{\mu'}^{fit})_3 \right) = 0,$$

$$g_{8+i} = (p_\gamma^{fit})_3 \left(v_i - (a_\gamma^{fit})_i \right) - (p_\gamma^{fit})_i \left(v_3 - (a_\gamma^{fit})_3 \right) = 0,$$

$\forall\, i \in \{1, 2\}$, while the index denotes Cartesian components of the three-vectors.

For each track two vertex constraints enter the system of equations. They are treated according to Sect. 4.6.2. Again, the initial state proton is not bound to the vertex for the same reason as in Sect. 5.1.2.

In case of the final state proton a special treatment is chosen to reflect the experimental situation of the CAMERA detector in the most adequate way. The role of the vertex constraint is taken by the interpolation constraints described in Sect. 4.6.3:

$$g_{10+i} = (p_{p'})_3\left((r_A)_i - (v)_i\right) - (p_{p'})_i\left((r_A)_3 - (v)_3\right) = 0,$$

$$g_{12+i} = (p_{p'})_3\left((r_B)_i - (v)_i\right) - (p_{p'})_i\left((r_B)_3 - (v)_3\right) = 0,$$

$\forall\, i \in \{1, 2\}$, while the index denotes Cartesian components of the three-vectors.

For each of the two hits, reconstructed in the inner and outer ring of the CAMERA detector, two extrapolation constraints enter the minimisation procedure.

In total 14 constraints are introduced into the procedure, while according to Eq. (6.13) seven free parameters have to be determined. Hence, the number of degrees of freedom is seven.

Looking at the difference of the longitudinal momentum between the initial beam and spectrometer measurement and the result of the kinematic fitting procedure, a shift is observed. It may be argued that at first order it is correct to compensate this shift by the kinematic fit, according to Fig. 6.12. On the other hand, the procedure is not designed to eliminate a bias on the measurement. The origin of this discrepancy between the measurement of the beam and the scattered muon is unknown. It is decided to modify the energy and momentum conservation constraints, to allow for a shift in the longitudinal momentum measurement:

$$\begin{aligned} g_3 &\approx -0.9\,\text{GeV/c}\ (-0.34\,\text{GeV/c for the Monte Carlo}), \\ g_4 &\approx -0.9\,\text{GeV}\ (-0.34\,\text{GeV for the Monte Carlo}), \end{aligned} \tag{6.15}$$

The influence on the results of Chap. 7 between Eqs. (6.14) and (6.15) is absorbed into the systematic error. This is demonstrated in Sect. 7.7.4.

Figures 6.13, 6.14 and 6.15 show the pull distributions of all input quantities with respect to the output quantities of the kinematic fitting procedure, taking Eq. (6.15) into account. The corresponding distributions without the division by the uncertainties are shown in Appendix A.4.4 inside Figs. A.34–A.36 and for an extended kinematic range inside Figs. A.40–A.42. The distributions without the application of Eq. (6.15) are shown inside Figs. A.37–A.39 and in case of an extended kinematic range inside Figs. A.43–A.45. Though the agreement between data and Monte Carlo is quite satisfactory, there are still visible deviations. This prevents the application of a single cut on the p-value of the kinematic fit, which would result in a far more elegant event selection.

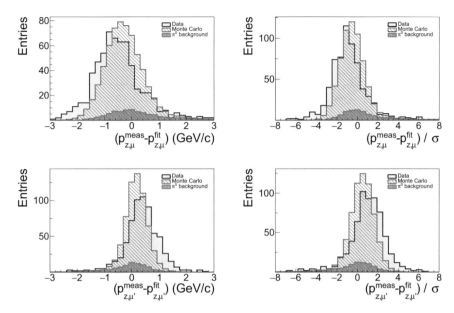

Fig. 6.12 Pull distributions of the longitudinal momentum of the incoming and outgoing muon, **using strict energy and momentum balance**: For better readability the abbreviations $p_{z,\mu}^{fit}$ and $p_{z,\mu'}^{fit}$ have been used for the longitudinal muon momenta, corrected by the kinematic fit. The measured longitudinal momenta of the in- and outgoing muon are denoted by $p_{z,\mu}^{meas}$ and respectively $p_{z,\mu'}^{meas}$ and are part of the track parameters defined in Eq. (4.7) of Sect. 4. The quantity σ is given by the respective elements of the in- and output covariance matrix by $\sigma = \sqrt{C_{5,5}^{meas} - C_{5,5}^{fit}}$, while the input covariance matrix is defined according to Eq. (4.8) of Sect. 4

Apart from the shift in the longitudinal momentum measurement the pull distributions are well centred around zero and show a slightly too large RMS value at the order of 1.2 in case of the data. Furthermore, it is interesting to see that for a single photon selection the π^0 background is almost not distinguishable from the signal.

As mentioned above, in case of the DVCS measurement the main advantage of the kinematic fit is to provide the most precise determination of t, the square of the four-momentum transfer to the proton. Figure 6.16 shows the achievable accuracy for $|t|$, given by the measurement of the CAMERA detector, a pure beam and spectrometer measurement and a combined measurement, making use of the kinematic fitting procedure. The values are extracted from a single photon Monte Carlo yield within a comparison with the generated values. The resolution of $|t|$ in case of the kinematic fitting is clearly improved compared to the two individual approaches. Especially for large values of $|t|$, where the resolution of the CAMERA detector gets worse quite rapidly, the spectrometer provides valuable information. The calculation of t by the beam and spectrometer measurement has been performed with Eq. (6.16), which is

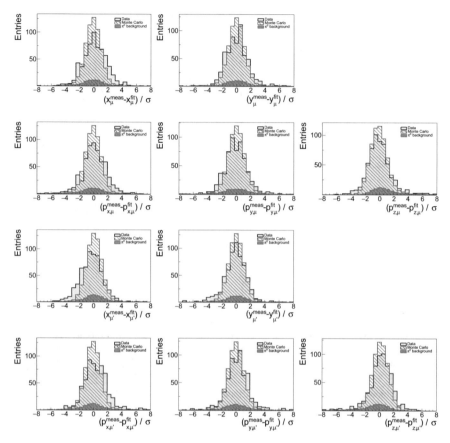

Fig. 6.13 Pull distributions of the track parameters for the in- and outgoing muon, **using Eq.** (6.15): The measured input track parameters of the incoming muon to the kinematic fitting procedure, defined according to Eq. (4.7) of Sect. 4.3, are denoted by $\left(x_\mu^{meas},\, y_\mu^{meas},\, p_{x,\mu}^{meas},\, p_{y,\mu}^{meas},\, p_{z,\mu}^{meas}\right)$ and the determined output parameters by $\left(x_\mu^{fit},\, y_\mu^{fit},\, p_{x,\mu}^{fit},\, p_{y,\mu}^{fit},\, p_{z,\mu}^{fit}\right)$. In case of the outgoing muon μ is replaced by μ'. The quantity σ is given by the respective elements of the in- and output covariance matrix C by $\sigma = \sqrt{C_{i,i}^{meas} - C_{i,i}^{fit}}$. The input covariance matrix is defined according to Eq. (4.8) of Sect. 4.3 and the index i satisfies $i \in \{1, \ldots, 5\}$

known as "constraint t" in the literature. It avoids the influence of the bad resolution of the measured photon energy on the determination of t. The derivation of Eq. (6.16) is shown in Appendix A.5.2.

$$t_{Spec.} = \frac{-Q^2 - 2(\nu/c)\left((\nu/c) - \sqrt{Q^2 + (\nu/c)^2}\cos\theta_{\gamma*\gamma}\right)}{1 + \frac{1}{m_p c^2}\left(\nu/c - \sqrt{Q^2 + (\nu/c)^2}\cos\theta_{\gamma*\gamma}\right)}. \tag{6.16}$$

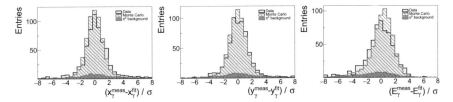

Fig. 6.14   Pull distributions of the track parameters of the photon, **using Eq.** 6.15: The measured input track parameters of the photon to the kinematic fitting procedure, defined according to Eq. (4.9) of Sect. 4.4, are the x- and y-positon of the reconstructed calorimeter cluster x_γ^{meas} and y_γ^{meas} and the reconstructed cluster energy E_γ^{meas}. The output parameters are denoted with the superscript "fit". The quantity σ is given by the respective elements of the in- and output covariance matrix C by $\sigma = \sqrt{C_{i,i}^{meas} - C_{i,i}^{fit}}$. The input covariance matrix is defined according to Eq. (4.10) of Sect. 4.4 and the index i satisfies $i \in \{1, \ldots, 3\}$

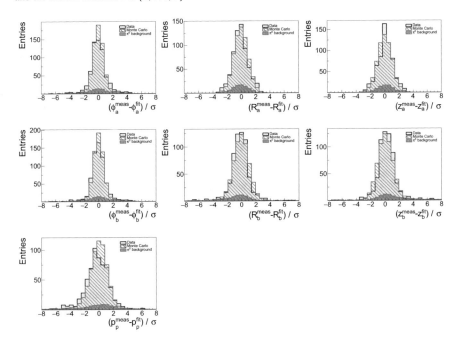

Fig. 6.15   Pull distributions of the proton track parameters, **using Eq.** 6.15: The measured input track parameters of the proton to the kinematic fitting procedure, defined according to Eq. (4.12) of Sect. 4.5, are given by $(\phi_{a,b}^{meas}, R_{a,b}^{meas}, z_{a,b}^{meas})$ the reconstructed hit positions in ring A and B and the magnitude of the reconstructed proton momentum p_p^{meas}. The output parameters are denoted with the superscript "fit". The quantity σ is given by the respective elements of the in- and output covariance matrix C by $\sigma = \sqrt{C_{i,i}^{meas} - C_{i,i}^{fit}}$. The input covariance matrix is defined according to Eq. (4.13) of Sect. 4.5 and the index i satisfies $i \in \{1, \ldots, 7\}$

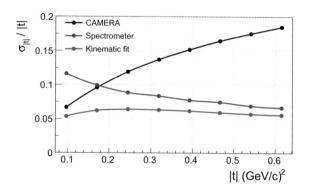

Fig. 6.16 Relative resolution on t, the square of the four-momentum transfer to the proton, as a function of $|t|$. The black line corresponds to a determination of t by using the CAMERA detector only. The blue line corresponds to a determination of t using the combined beam and spectrometer measurement of the in- and outgoing muon, according to Eq. (6.16). The red line shows the most accurate determination of t by combining the beam and spectrometer measurement with the CAMERA measurement, using the kinematic fitting procedure. The resolutions have been extracted by comparing reconstructed and generated values of $|t|$, using a single photon Monte Carlo yield

References

1. G. Ingelman, A. Edin, J. Rathsman, LEPTO 6.5: a Monte Carlo generator for deep inelastic lepton - nucleon scattering. Comput. Phys. Commun. **101**, 108–134 (1997). https://doi.org/10.1016/S0010-4655(96)00157-9
2. C. Regali, Exclusive event generation for the COMPASS II experiment at CERN and improvements for the Monte-Carlo chain, Dissertation. Albert Ludwigs Universität Freiburg (2016). https://doi.org/10.6094/UNIFR/11449, https://na58-project-tgeant.web.cern.ch/
3. P.A.M. Guichon, N. d'Hose, A. Vidon, Calculation of the pure Bethe Heitler cross section (2015). Private communications
4. L.L. Frankfurt, A. Freund, M. Strikman, Diffractive exclusive photon production in DIS at DESY HERA. Phys. Rev. D**58** (1998). https://doi.org/10.1103/PhysRevD.58.114001, [Erratum: Phys. Rev. D**59** 119901 (1999)]
5. L.L. Frankfurt, A. Freund, M. Strikman, Deeply virtual compton scattering at HERA A probe of asymptotia. Phys. Lett. B **460**, 417–424 (1999). https://doi.org/10.1016/S0370-2693(99)00803-5
6. A. Sandacz, Modifications to FFS model and predictions, https://www.compass.cern.ch/compass/gpd/meetings/200904_april22/AS_gpd-Apr2009.ppt
7. A.V. Belitsky, D. Mueller, A. Kirchner, Theory of deeply virtual Compton scattering on the nucleon. Nucl. Phys. B **629**, 323–392 (2002). https://doi.org/10.1016/S0550-3213(02)00144-X
8. L.W. Mo, Y.S. Tsai, Radiative corrections to elastic and inelastic e p and μ p Scattering. Rev. Mod. Phys. **41**, 205–235 (1969). https://doi.org/10.1103/RevModPhys.41.205
9. S.V. Goloskokov, P. Kroll, Transversity in hard exclusive electroproduction of pseudoscalar mesons. Eur. Phys. J. A **47**, 112 (2011). https://doi.org/10.1140/epja/i2011-11112-6
10. T. Szameitat, New Geant4-based Monte Carlo Software for the COMPASS-II Experiment at CERN, Dissertation. Albert Ludwigs Universität Freiburg (2017), https://doi.org/10.6094/UNIFR/11686, https://na58-project-tgeant.web.cern.ch/
11. A. Gross, Extraction of cross sections for Rho and Phi muoproduction at the COMPASS experiment, Master thesis, Albert Ludwigs Universität Freiburg (2014)

12. A. Ferrero, Extraction of the target position from the 2012 DVCS data (2014). Private communications
13. E. Fuchey, Extraction of the cluster time with respect to the trigger as a function of the cluster energy (2014). Private communications

Chapter 7
The Cross Section and Its t-Dependence

During this chapter the DVCS cross section and its exponential t-dependence are extracted. The exponential t-dependence is denoted as the t-slope or simply by the symbol B in the following. The first part of the chapter describes the cross section extraction method, the background estimation, the normalisation of the Bethe–Heitler contribution and the acceptance correction. Finally, the extracted DVCS cross section is presented and the t-slope is determined. Within the second part of the chapter systematic uncertainties on the measurement are discussed. The chapter concludes with an interpretation of the results.

7.1 Extraction Method for the DVCS Cross Section

The aim is to extract the t-dependence of the pure DVCS cross section of the process:

$$\gamma^* p \to \gamma p',$$

from count rates of the process:

$$\mu p \to \mu' p' \gamma,$$

in the kinematic range:

$$0.08\,(\text{GeV/c})^2 < |t| < 0.64\,(\text{GeV/c})^2,$$

$$1\,(\text{GeV/c})^2 < Q^2 < 5\,(\text{GeV/c})^2,$$

$$10\,\text{GeV} < \nu < 32\,\text{GeV}.$$

© Springer International Publishing AG, part of Springer Nature 2018
P. Jörg, *Exploring the Size of the Proton*, Springer Theses,
https://doi.org/10.1007/978-3-319-90290-6_7

Table 7.1 The four bins in t

Bin	t_1	t_2	t_3	t_4
Range in $(\text{GeV/c})^2$	]0.08, 0.22]	]0.22, 0.36]	]0.36, 0.5]	]0.5, 0.64[

Table 7.2 Bins in Q^2 and ν

Bin		Q_1^2	Q_2^2	Q_3^2	Q_4^2	
Range in $(\text{GeV/c})^2$		]1, 2]	]2, 3]	]3, 4]	]4, 5[	
Bin	ν_1	ν_2	ν_3	...	ν_9	ν_{10}	ν_{11}
Range in GeV	]10, 12]	]12, 14]	]14, 16]	...	]26, 28]	]28, 30]	]30, 32[

Four bins in $|t|$, according to Table 7.1, are used. The mean cross section in these four bins is constructed as follows:

$$\left\langle \frac{d\sigma_{\text{DVCS}}^{\gamma^* p \to \gamma p'}}{d|t|} \right\rangle_n^{\pm} = \frac{\sum_{ij} \left(\frac{d\sigma_{\text{DVCS}}^{\gamma^* p \to \gamma p'}}{d|t|} \right)_{ijn}^{\pm} \Delta Q_i^2 \Delta \nu_j}{\sum_i \Delta Q_i^2 \sum_j \Delta \nu_j}. \tag{7.1}$$

Here n denotes the index for the bin in $|t|$, i the index for the bin in Q^2, j the index for the bin in ν and $\pm$ the beam charge. Equation (7.1) states that the average differential cross section in each of the four bins in t is given as a weighted mean over the differential cross sections extracted in bins of Q^2 and ν, according to Table 7.2.

Since it is necessary to correct the data for the Bethe–Heitler contribution (BH) and a possible π^0 contamination, the differential cross section of the process $\mu p \to \mu' \gamma p'$ in a certain bin of t, Q^2 and ν is given by the following relation:

$$\left\langle \frac{d\sigma_{\text{DVCS}}^{\mu p \to \mu' \gamma p'}}{d|t| d Q^2 d\nu} \right\rangle_{ijn}^{\pm} = \left\langle \frac{d\sigma_{\text{data}}^{\mu p \to \mu' \gamma p'}}{d|t| d Q^2 d\nu} \right\rangle_{ijn}^{\pm} - \left\langle \frac{d\sigma_{\text{BH}}^{\mu p \to \mu' \gamma p'}}{d|t| d Q^2 d\nu} \right\rangle_{ijn}^{\pm} - \left\langle \frac{d\sigma_{\pi^0}^{\mu p \to \mu' \gamma p'}}{d|t| d Q^2 d\nu} \right\rangle_{ijn}^{\pm}. \tag{7.2}$$

However, to extract a cross section for virtual-photon proton scattering from muon proton scattering, relation (7.3) is used. It contains the transverse virtual-photon flux $\Gamma(Q^2, \nu)$:

$$\left\langle \frac{d\sigma_{\text{DVCS}}^{\gamma^* p \to \gamma p'}}{d|t|} \right\rangle^{\pm} = \left\langle \frac{1}{\Gamma(Q^2, \nu)} \frac{d\sigma_{\text{DVCS}}^{\mu p \to \mu' \gamma p'}}{d|t| d Q^2 d\nu} \right\rangle^{\pm}, \tag{7.3}$$

while[1]:

---

[1]Replacing the convention dependent factor within Ref. [1] by $k = \nu(1 - x_{\text{Bj}})$, according to the Hand convention [2], yields the quoted expression.

$$\Gamma(Q^2, \nu) = \frac{\alpha_{em}(1 - x_{Bj})}{2\pi Q^2 y E} \left[y^2 \left(1 - \frac{2m_\mu^2}{Q^2}\right) + \frac{2}{1 + \left(\frac{Q^2}{\nu^2}\right)} \left(1 - y - \frac{Q^2}{4E^2}\right) \right],$$

according to the Hand convention. Using Eq. (7.2) together with (7.3) results in:

$$\left\langle \frac{d\sigma_{DVCS}^{\gamma^* p \to \gamma p'}}{d|t|} \right\rangle_{ijn}^{\pm} = \left\langle \frac{1}{\Gamma} \frac{d\sigma_{data}^{\mu p \to \mu' \gamma p'}}{d|t| d Q^2 d\nu} \right\rangle_{ijn}^{\pm} - \left\langle \frac{1}{\Gamma} \frac{d\sigma_{BH}^{\mu p \to \mu' \gamma p'}}{d|t| d Q^2 d\nu} \right\rangle_{ijn}^{\pm} - \left\langle \frac{1}{\Gamma} \frac{d\sigma_{\pi^0}^{\mu p \to \mu' \gamma p'}}{d|t| d Q^2 d\nu} \right\rangle_{ijn}^{\pm}.$$

Transforming this equation one can see how the acceptance enters and what is technically done during the extraction procedure:

$$\left\langle \frac{d\sigma_{DVCS}^{\gamma^* p \to \gamma p'}}{d|t|} \right\rangle_{ijn}^{\pm} = \frac{(a_{ijn}^{\pm})^{-1}}{\mathcal{L}^{\pm} \Delta t_n \Delta Q_i^2 \Delta \nu_j} \left(\sum_{e=1}^{N_{ijn}^{data,\pm}} \frac{1}{\Gamma(Q_e^2, \nu_e)} \right.$$

$$\left. - c_{BH}^{\pm} \sum_{e=1}^{N_{ijn}^{BH,\pm}} \frac{(w_{P.A.M})_e}{\Gamma(Q_e^2, \nu_e)} - c_{\pi^0}^{\pm} \sum_{e=1}^{N_{ijn}^{\pi_\gamma^0,\pm}} \frac{(w_{\pi^0})_e}{\Gamma(Q_e^2, \nu_e)} \right).$$

(7.4)

The intermediate steps are displayed in Sect. A.5.3. The first term in Eq. (7.4) states that one has to sum the factor $\frac{1}{\Gamma(Q_e^2,\nu_e)}$ of each event inside the bin (ijn) and divide this sum by the bin width $(\Delta t_n \Delta Q_i^2 \Delta \nu_j)$ corrected by the acceptance $a_{ijn}^{\pm}$ times the luminosity $\mathcal{L}^{\pm}$. In this sense $\Gamma(Q_e^2, \nu_e)$ can be regarded as a weight for each event e or in other words as an event by event kinematic pre-factor.

The last two terms in Eq. (7.4) are estimated by Monte Carlo. The number of events inside the bin (ijn) are denoted as $N_{ijn}^{data,\pm}$ for the data, $N_{ijn}^{BH,\pm}$ for the Bethe–Heitler Monte Carlo and $N_{ijn}^{\pi_\gamma^0,\pm}$ for the π^0 Monte Carlo. The factors $c_{BH}^{\pm}$ and $c_{\pi^0}^{\pm}$ account for the correct normalisation of the Monte Carlos to the measured data. The normalisation of the Bethe–Heitler Monte Carlo will be described in Sect. 7.3. The event weight of the π^0 Monte Carlo is generically denoted as w_{π^0}. It accounts for two different types of π^0 background Monte Carlos generated by LEPTO and HEPGen++. The estimation of the π^0 background is the topic of the following section.

Inserting Eq. (7.4) into (7.1) results in:

$$\left\langle \frac{d\sigma_{DVCS}^{\gamma^* p \to \gamma p'}}{d|t|} \right\rangle_n^{\pm} = \frac{1}{\mathcal{L} \Delta t_n \Delta Q^2 \Delta \nu} \sum_{ij} \left[(a_{ijn}^{\pm})^{-1} \left(\sum_{e=1}^{N_{ijn}^{data,\pm}} \frac{1}{\Gamma(Q_e^2, \nu_e)} \right.\right.$$

$$\left.\left. - c_{BH}^{\pm} \sum_{e=1}^{N_{ijn}^{BH,\pm}} \frac{(w_{P.A.M})_e}{\Gamma(Q_e^2, \nu_e)} - c_{\pi^0}^{\pm} \sum_{e=1}^{N_{ijn}^{\pi_\gamma^0,\pm}} \frac{(w_{\pi^0})_e}{\Gamma(Q_e^2, \nu_e)} \right) \right].$$

(7.5)

Here $\Delta Q^2 = \sum_i \Delta Q_i^2 = 4\,(\text{GeV}/c)^2$ and $\Delta\nu = \sum_j \Delta\nu_j = 22\,\text{GeV}$ denote the total width of the extraction regime in Q^2 and ν. Finally, the contribution of both muon charges are summed:

$$\left\langle \frac{d\sigma_{\text{DVCS}}^{\gamma^* p \to \gamma p'}}{d|t|} \right\rangle_n = \frac{1}{2}\left(\left\langle \frac{d\sigma_{\text{DVCS}}^{\gamma^* p \to \gamma p'}}{d|t|} \right\rangle_n^+ + \left\langle \frac{d\sigma_{\text{DVCS}}^{\gamma^* p \to \gamma p'}}{d|t|} \right\rangle_n^- \right). \tag{7.6}$$

Equation (7.6) represents the differential cross section in the nth bin of $|t|$.

7.2   Estimation of the π^0 Background

The production of π^0, which decay into two photons, is the major background source for a detection of exclusive single photon production. Two cases have to be distinguished:

- Additional photons of π^0 decays are detected in the electromagnetic calorimeters which have an energy below the thresholds used for the single photon reconstruction (see Sect. 6.2.2). This will be denoted as the $\pi^0_{\gamma\gamma}$ background contribution and will be discussed in Sect. 7.2.1.
- The additional photon of a π^0 decay could escape detection. This contribution shall be denoted as the π^0_γ background contribution. It is estimated by Monte Carlo techniques as it will be described in Sects. 7.2.1 and 7.2.2.

In both cases the energy of the additional photon is rather low. Otherwise, the polluted events would not have passed the exclusivity cuts of Sect. 6.2.3.

Neither the semi-inclusive π^0 production cross section close to $z = E_{\pi^0}/\nu = 1$ nor the exclusive π^0 production cross section are well constrained within the kinematical region of COMPASS-II. Thus, data driven methods together with Monte Carlo predictions have to be used to estimate the π^0 contamination. These methods will be described in the following two sections.

7.2.1   The $\pi^0_{\gamma\gamma}$ and π^0_γ Background

The $\pi^0_{\gamma\gamma}$ background contribution to the single photon sample of Sect. 6.2 can be directly identified within the data. For each event of the final sample photon pairs are created by combining all additionally detected photons with the single photon. Figure 7.1 shows the mass distribution of these photon pairs, separated for the overall and the two data yields of different beam charge. A clear peak at the nominal π^0 mass is visible. The events within this peak comprise the $\pi^0_{\gamma\gamma}$ background. As it was already indicated in Table 6.1, these events are rejected from the final sample by applying the cut:

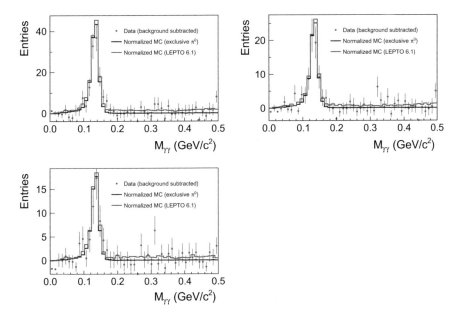

Fig. 7.1 Invariant mass of the two γ system: The DVCS photon is combined with all other photons below the DVCS energy thresholds detected in ECal0 or ECal1. The LEPTO and HEPGen++ Monte Carlos are individually normalised to the amount of visible leaking π^0 in the data. The HEPGen++ Monte Carlo is denoted with the term "(exclusive π^0)" within the plot. ECal2 is excluded from the selection since there is no visible π^0 mass peak. Top Left: Combined μ^+ and μ^- data yield. Top Right: μ^- data yield. Bottom Left: μ^+ data yield

$$|m_{\gamma\gamma} - m_{\pi^0}| > 20\,\text{MeV}/c^2, \tag{7.7}$$

while m_{π^0} denotes the nominal mass of the π^0.

However, the very same events are also used to estimate the amount of π^0_γ background. Therefore, the two different Monte Carlo yields of Fig. 7.1 are used. They are normalised to the observed $\pi^0_{\gamma\gamma}$ yield in the data. In order to increase the statistical robustness of this normalisation, the kinematic range for the detection of the $\pi^0_{\gamma\gamma}$ contribution is given by:

$$1\,(\text{GeV}/c)^2 < Q^2 < 20\,(\text{GeV}/c)^2, 0.08\,(\text{GeV}/c)^2 < |t| < 0.64\,(\text{GeV}/c)^2$$

$$\text{and } 8\,\text{GeV} < \nu < 144\,\text{GeV}.$$

In Fig. 7.1 the LEPTO Monte Carlo yield is shown in blue. It accounts for the contribution of semi-inclusive π^0 production. The Monte Carlo yield used to estimate the contribution of exclusive π^0 production is shown in black. It is produced by the event generator HEPGen++, using the event weight w_{π^0} of Sect. 6.1.1.

The estimation of the π_γ^0 background relies on the HepGen++ and the LEPTO Monte Carlos. As one applies the event selection of Sect. 6[2] to the two normalised Monte Carlos, the π_γ^0 background contribution of each Monte Carlo is given by the remaining yields. However, before one can use the two Monte Carlos to correct the data, it has to be clarified which amount of the $\pi_{\gamma\gamma}^0$ contribution in the data is given by either of the two Monte Carlo predictions. It is clear that they can not both be normalised to the observed $\pi_{\gamma\gamma}^0$ contribution in the data, which would lead to double counting of the estimated π_γ^0 yield. This is taken into account by the parameter r_H. It describes the contribution of the HEPGen++ Monte Carlo to the $\pi_{\gamma\gamma}^0$ background. The last term in Eq. (7.4) can thus be written as:

$$
c_{\pi_\gamma^0}^{\pm} \sum_{e=1}^{N_{ijn}^{\pi_\gamma^0,\pm}} \frac{(w_{\pi^0})_e}{\Gamma(Q_e^2, \nu_e)} = c_H^{\pm}(r_H) \sum_{e=1}^{N_{ijn}^{H,\pm}} \frac{(w_{\pi^0})_e}{\Gamma(Q_e^2, \nu_e)} + c_L^{\pm}(1 - r_H) \sum_{e=1}^{N_{ijn}^{L,\pm}} \frac{1}{\Gamma(Q_e^2, \nu_e)} \quad (7.8)
$$

The normalisations $c_H^{\pm}$ of the HEPGen++ and $c_L^{\pm}$ of the LEPTO Monte Carlos to the observed $\pi_{\gamma\gamma}^0$ yield in the data are taken as illustrated in Fig. 7.1, while the number of events of the two Monte Carlos are denoted by $N_{ijn}^{H,\pm}$ and $N_{ijn}^{L,\pm}$. The estimation of the parameter r_H is the topic of the next section.

7.2.2 Normalisation of the LEPTO and HEPGen++ π^0 Monte Carlos

For the estimation of the parameter r_H the event selection of Table 7.3 is used. The selection is optimised to select exclusive π^0 events and is described in detail in Ref. [3].

Figure 7.2 shows the invariant mass of the photon pairs, remaining after the event selection. The LEPTO Monte Carlo is shown in green, while the HEPGen++ Monte Carlo is shown in blue. Both Monte Carlos are normalised to the number of detected π^0 events in the data within the peak around the nominal π^0 mass. The peak region is indicated by the red lines.

The basic idea to separate the contributions of the two Monte Carlos relies on the shape of the distributions of the exclusivity variables outlined in Table 7.3.[3] These variables are particular sensitive to semi-inclusive background. The procedure goes as follows: The normalisation shown in Fig. 7.2 is used, while one of the exclusivity cuts is removed from the event selection and finally the cut shown in Fig. 7.2 is applied. The distribution of the removed exclusivity variable for the data is compared to the distributions of the two Monte Carlo yields. Denoting the three distributions

---

[2]This includes in particular Eq. (7.7), which removes the $\pi_{\gamma\gamma}^0$ contribution from the samples.

[3]The exclusivity variables can be found within the block called "exclusivity selections".

Table 7.3 Overview of the selection of exclusive π^0 events

Events with:	General event criteria

• Time in spill: $1\,\text{s} < T < 10.4\,\text{s}$
• Considered trigger types:
     Middle Trigger (MT) or
     Ladder Trigger (LT) or
     Outer Trigger (OT)	

Primary vertices with:
• Vertex z-position: $-311.2\,\text{cm} < v_z < -71.2\,\text{cm}$
• Vertex distance from target centre: $d < 1.9\,\text{cm}$
   (see Sect. 6.2.1)
• One incoming charged track, μ with:
   >2 hits in the Beam Momentum Stations (BMS),
   >1 hit in the Scintillating Fibre detectors (Fi),
   >2 hits in the Silicon detectors (Si),
 beam momentum: $140\,\text{GeV/c} < p_\mu < 180$ GeV/c, beam track traverses the full target volume *(see Sect. 6.2.1)*
• One outgoing charged track μ' with:
   same charge than incoming track,
   traversed radiation lengths: $X/X_0 > 15$,
   z-position of first measured point: $z_{first} < 350\,\text{cm}$,
   z-position of last measured point: $z_{last} > 350\,\text{cm}$
• Inclusive scattering variables:
   energy loss: $\nu > 8\,\text{GeV}$,
   photon virtuality: $Q^2 > 1\,(\text{GeV/c})^2$
	(1) Muon and vertex selection

Neutral clusters γ_i with:
• A detection in ECal0 or ECal1
• A valid cluster time
   (see Sect. 6.2.2)
• A reconstructed cluster energy: $E(\gamma_i) > 0.3\,\text{GeV}$, $\forall i$
• $\exists\, l : E(\gamma_l) > 1\,\text{Gev}$ for Ecal0
   or $\exists\, l : E(\gamma_l) > 2\,\text{Gev}$ for Ecal1
	(2) Photon selection

Reconstructed CAMERA tracks with:
• Longitudinal hit position z inside ring A and B:
   $-366.19\,\text{cm} < z_A < 8.81\,\text{cm}$,
   $-338.94\,\text{cm} < z_B < 71.06\,\text{cm}$
• Velocity of reconstructed recoiling particle:
   $0.1 < \beta := \frac{v}{c} < 1$
	(3) CAMERA selections

(continued)

Table 7.3 (continued)

All combinations of (1), (2) and (3) which satisfy:	Exclusivity selections
• $\lvert \Delta p_T \rvert < 0.3\,\text{GeV/c}$ • $\lvert \Delta \phi \rvert < 0.4\,\text{rad}$ • $\lvert \Delta z_A \rvert < 16\,\text{cm}$ • $\lvert M_X^2 \rvert < 0.3\,\left(\text{GeV/c}^2\right)^2$ *(see Sect. 6.2.3 for the definitions,* *replace p_γ with $(p_{\gamma,i} + p_{\gamma,j})$ for $i \neq j$)* • Exactly one combination must be left	

Fig. 7.2 Invariant mass of the two γ system after the application of the cuts described in Sect. 7.2.2. The applied cut in order to select the π^0 contribution is shown in red. The exclusive π^0 Monte Carlo (HepGen++) and the LEPTO Monte Carlo have been normalised to the data

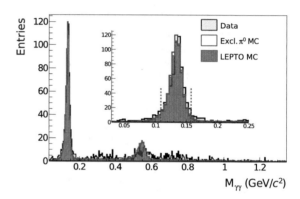

representing the data, the LEPTO and the HEPGen++ Monte Carlo by the set V, a least squares function χ^2 is constructed as follows:

$$\chi_V^2(a, b) = \sum_{i=1}^{N_{bins}^V} \frac{\left(N_i^D - a N_i^H - b N_i^L\right)^2}{\left(\sigma_i^D\right)^2 + \left(a \sigma_i^H\right)^2 + \left(b \sigma_i^L\right)^2}. \tag{7.9}$$

The set V is explicitly given by $V = \{\vec{N}^D, \vec{\sigma}^D, \vec{N}^H, \vec{\sigma}^H, \vec{N}^L, \vec{\sigma}^L\}$. The bin contents of the respective distributions and their statistical uncertainties are depicted by $\vec{N}^{D;L;H}$ and $\vec{\sigma}^{D;L;H}$ for the data (D), the LEPTO (L) and the HEPGen++ (H) Monte Carlos, while N_{bins}^V denotes the number of bins. The parameters for which the least squares function will be minimised are given by a and b. They describe the contribution of the two Monte Carlo yields in order to fit the data best. In particular, the following three methods are applied to the exclusivity distributions:

Method 1:

The parameters a and b are chosen according to Eq. (7.10), while r_H denotes the contribution of the HEPGen++ Monte Carlo and the index S the set of the three distributions of the respective exclusivity variable:

$$a = r_H, \quad b = (1 - r_H). \tag{7.10}$$

Hence, a single parameter r_H is used in order to describe the data:

$$\chi^2(r_H) := \chi^2_S(r_H, (1 - r_H)) \text{ (using } V = S \text{ in Eq. 6.9).}$$

Figure 7.3 shows the fit result for the distribution of the undetected mass squared. The data is shown in yellow, the sum of the two Monte Carlo yields in red and the HEPGen++ Monte Carlo in blue.

Method 2:

In order to gain confidence in the estimate given by the first method, a different approach uses in addition to S the set of background like distributions B. The set B denotes the three distributions (data, LEPTO, HEPGen++) of a exclusivity variable in case there is more than one π^0 candidate[4] left after applying the event selection. These events are most likely of semi-inclusive origin and make the background like set S particular sensitive on the contribution of the LEPTO Monte Carlo. Surely, in this case the very last cut of Table 7.3 has to be removed.

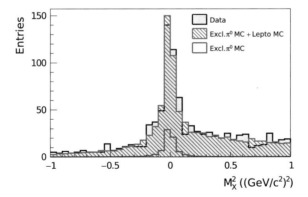

Fig. 7.3 Distribution of M_X^2 for **Method 1** of Sect. 7.2.2. The blue histogram describes the overall Monte Carlo estimate given by the exclusive π^0 (HEPGen++) and the LEPTO Monte Carlo yields, while the red histogram displays the fraction described by the exclusive π^0 Monte Carlo yield

---

[4]The π^0 candidates are denoted as combinations at the very bottom of Table 7.3.

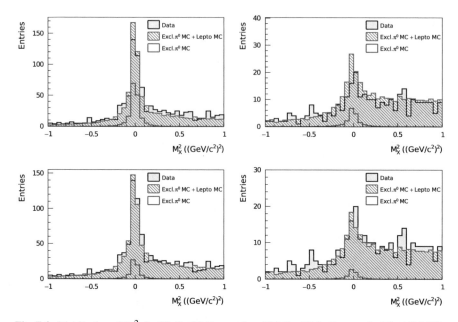

Fig. 7.4 Distribution of M_X^2 for **Method 2** (top row) and **Method 3** (bottom row) of Sect. 7.2.2 for $N_B < 3$. The blue histogram describes the overall Monte Carlo estimate given by the exclusive π^0 (HEPGen++) and the LEPTO Monte Carlo yields, while the red histogram displays the fraction described by the exclusive π^0 Monte Carlo yield. Left: Set of signal distributions S. Right: Set of background like distributions B

The least squares function is build from the two sets S and B. It has the following form:

$$\chi^2(r_H) := \chi_S^2(r_H, (1 - r_H)) + \chi_B^2(r_H, (1 - r_H)).$$

For the background like distributions a further distinction has to be made. As the number of final π^0 candidates is greater than one, it must be at least one after the π^0 mass region of Fig. 7.2 has been selected. The number of π^0 candidates after the event selection and the final mass selection is denoted as the background multiplicity N_B. The top row of Fig. 7.4 shows the fit result in the same fashion as for Method 1 in case of $N_B < 3$. One can observe that the agreement between data and Monte Carlo is quite unsatisfactory.

Method 3:

Method 3 is almost similar to Method 2 apart from the fact that in this case the χ^2 function depends on two parameters. In addition to the parameter r_H a second parameter r_L^B is introduced:

$$\chi^2(r_H, r_L^B) := \chi_S^2(r_H, (1 - r_H)) + \chi_B^2(r_H, r_L^B).$$

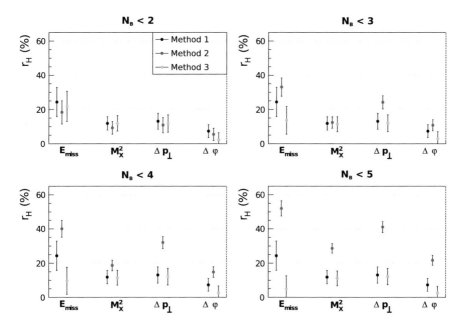

Fig. 7.5 Comparison of the fraction r_H for the three methods described in Sect. 7.2.2 for different background multiplicities N_B after the cut on the π^0 mass and different background sensitive variables. The variables are explained in Table 7.3, apart from E_{miss}. The quantity $E_{miss} := \frac{(p+q-p_{\pi^0})^2 - M_p^2}{2M_p}$ is the missing energy, while M_p denotes the mass of the proton and p, q, p_{π^0} the four-momenta of the initial proton, the virtual photon and the π^0 candidate

The idea behind is that the LEPTO Monte Carlo is most likely not providing a good absolute description of the background multiplicity N_B and thus a second parameter is needed for the normalisation of the background like LEPTO Monte Carlo distributions. It should be emphasised, that in case of both sets of distributions the same parameter for the HEPGen++ Monte Carlo yield is used. The bottom row of Fig. 7.4 shows the fit result in the same fashion as for Method 2.

A comparison of the results of the three methods is shown in Fig. 7.5 for different values of the background multiplicity N_B and different exclusivity variables. The distributions of the remaining exclusivity variables in case of $N_B < 3$ are shown in Appendix A.5.1. One clearly observes that in case of Method 2 the resulting value of r_H depends strongly on the multiplicity requirement N_B. This fact was already remarked above where it served as the motivation of Method 3.

Considering the results of Method 1 and Method 3, the value of r_H in Eq. (7.8) is chosen to be at the order of $r_H = 0.1$, while a systematic uncertainty of 0.2 is considered in Sect. 7.6.2. For completeness it should be noted that this estimate of r_H takes also into account that the event selection of the $\pi_{\gamma\gamma}^0$ contribution of Sect. 7.2.1

differs from the event selection of this section. As one applies the selection of the $\pi^0_{\gamma\gamma}$ contribution to the two normalised Monte Carlo yields of this section, a slight increase of r_H at the order of one percent is observed.

7.3 Normalisation of the Bethe–Heitler Contribution

As stated in Sect. 7.1, the Bethe–Heitler yield needs to be subtracted from the data inside the extraction region of the DVCS cross section. This is achieved by Monte Carlo. A single photon Monte Carlo sample with the weight $w_{P.A.M.}$[5] is used. It is normalised to the luminosity of the 2012 data according to Eq. (6.6) and shall be denoted as the Bethe–Heitler Monte Carlo in the following.

In order to estimate systematic effects on the cross section measurement, the Bethe–Heitler Monte Carlo together with the π_γ background Monte Carlo are used. Both Monte Carlo predictions are compared to the measured data as a function of $\phi_{\gamma*\gamma}$[6] in three kinematically different regions:

- The "reference region" of (80 GeV $< \nu <$ 144 GeV):
  In this region the Bethe–Heitler process completely dominates the exclusive single photon yield while a negligible π^0 contamination is estimated. The data yield is supposed to agree with the hypothesis of a pure Bethe–Heitler contribution on the percent level. Thus, this region is supposed to be described by the Bethe–Heitler Monte Carlo only.
- The "interference region" of (32 GeV $< \nu <$ 80 GeV):
  For this region the Bethe–Heitler cross section is still dominant, but the DVCS contribution becomes sizeable and is boosted by the interference term, as described in Sect. 2.4. Due to the interference between the DVCS and the Bethe–Heitler process a slight asymmetry of the $\phi_{\gamma*\gamma}$ distribution is expected.
- The "DVCS extraction region" of (10 GeV $< \nu <$ 32 GeV):
  In this region the DVCS amplitude is sizeable. A significant difference between the sum of the Bethe–Heitler and the π^0_γ Monte Carlo in contrast to the extracted amount of single photon events in the data is expected.

The comparison of data and Monte Carlo for the three regions is shown in Fig. 7.6 separately for the data yields taken with the μ^- and the μ^+ beam and the sum of both.

For the "reference region" one clearly observes that there is a loss of Bethe–Heitler events at the order of 20% in the data taken with μ^+ beam and an excess of about 10% for the μ^- beam compared to the Bethe–Heitler Monte Carlo yields. This discrepancy might get smaller or vanish for small values of ν, but unfortunately there exists no reference yield in this region. Despite all efforts the source of the uncertainty in

---

[5]The weight $w_{P.A.M.}$ represents the Bethe–Heitler calculation including the muon mass in the propagator (see Sect. 6.1).

[6]For the definition of $\phi_{\gamma*\gamma}$ see Fig. 2.13.

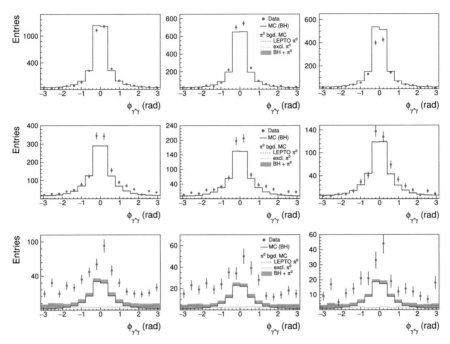

Fig. 7.6 Distributions of $\phi_{\gamma^*\gamma}$ for $\left(0.08\,(\text{GeV/c})^2 < |t| < 0.64\,(\text{GeV/c})^2\right)$ in three regions of ν. Top row: The "reference region" of $\left(80\,\text{GeV} < \nu < 144\,\text{GeV}\right)$. Middle row: The "interference region" of $\left(32\,\text{GeV} < \nu < 80\,\text{GeV}\right)$. Bottom row: The "DVCS extraction region" of $\left(10\,\text{GeV} < \nu < 32\,\text{GeV}\right)$. The Bethe–Heitler Monte Carlo (MC (BH)) has been normalised to the total integrated luminosity of the 2012 data for the overall μ^+ and μ^- data yield (left column), the μ^- data yield (middle column) and the μ^+ data yield (right column). The π^0_γ background (π^0 bgd. MC) is estimated according to Sect. 7.2. The sum of the Bethe–Heitler Monte Carlo and the π^0_γ background estimate (BH + π^0) is shown in blue. The data and either of the Monte Carlo yields are not corrected for acceptance effects. The top and middle row correspond to $\left(1\,(\text{GeV/c})^2 < Q^2 < 20\,(\text{GeV/c})^2\right)$, while the bottom row corresponds to $\left(1\,(\text{GeV/c})^2 < Q^2 < 5\,(\text{GeV/c})^2\right)$

the "reference region" of large ν is unknown. Therefore, a conservative approach is chosen, to absorb the influence of an equally large discrepancy inside the "extraction region" of small ν into the systematic error on the cross section and its t-dependence. This approach is detailed in Sect. 7.6.1.

7.4 Acceptance Corrections

The acceptance correction factors are extracted by applying the event selection of Sect. 6.2 to a single photon Monte Carlo sample using the event generator HEPGen++ and the DVCS event weight w_{DVCS}. Denoting $N_g(\Delta\Omega)$ the sum of DVCS weights of the generated Monte Carlo events in the phase space $\Delta\Omega$ and $N_r(\Delta\Omega_r)$ the sum of reconstructed DVCS event weights, the acceptance correction factor a for the phase space element $\Delta\Omega_r$ is given by:

$$a(\Delta\Omega_r) = \frac{N_r(\Delta\Omega_r)}{N_g(\Delta\Omega)}.$$

The index r emphasises the fact that the quantity $N_r(\Delta\Omega_r)$ is increased by the respective event weight in case the values of the reconstructed kinematic variables are found to be within the phase space element Ω_r. Hence, this definition of the acceptance is also taking into account kinematical smearing effects, due to the experimental resolution on the reconstructed kinematic variables.[7]

Figure 7.7 shows the acceptance for the DVCS process as a function of Q^2, ν and $\phi_{\gamma*\gamma}$ individually for both beam charges. One can clearly observe that for the region of $(8\,\text{GeV} < \nu < 10\,\text{GeV})$ and $(Q^2 > 3\,(\text{GeV/c})^2)$ and for the region of $(10\,\text{GeV} < \nu < 12\,\text{GeV})$ and $(Q^2 > 5\,(\text{GeV/c})^2)$ the acceptance tends to drop to zero as one approaches $\phi_{\gamma*\gamma}$ close to $\pm\pi$. However, for the extraction of the DVCS cross section a single bin in Q^2 and ν, avoiding regions of zero acceptance, has to be chosen. This is why the extraction region is limited to:

$$(10\,\text{GeV} < \nu < 32\,\text{GeV}) \text{ and } (1\,(\text{GeV/c})^2 < Q^2 < 5\,(\text{GeV/c})^2).$$

Figure 7.8 shows the acceptance as a function of Q^2, $|t|$ and $\phi_{\gamma*\gamma}$. A rather flat and symmetric behaviour with respect to $\phi_{\gamma*\gamma} = 0$ of the acceptance as a function of $\phi_{\gamma*\gamma}$ is observed. According to Eq. (2.31) the interference term between the DVCS and the Bethe–Heitler process is odd with respect to $\phi_{\gamma*\gamma}$. Hence, it cancels naturally for an even acceptance in $\phi_{\gamma*\gamma}$, without parametrising the acceptance as a function of $\phi_{\gamma*\gamma}$.

In order to make the best use of the available Monte Carlo statistics, it is thus decided to parametrise the acceptance for the extraction of the DVCS cross section as a function of $|t|$, Q^2 and ν as shown in Fig. 7.9.[8] This particular choice of the acceptance binning is motivated by the dependence of the transverse virtual photon flux on Q^2 and ν, as it is described in Sect. 7.1.

---

[7] Within Figs. 7.7, 7.8 and 7.9 the kinematic quantities, determined by the kinematic fit, are used.

[8] With this choice of the acceptance parametrisation, the $\phi_{\gamma*\gamma}$ modulations of the coefficients c_1^{DVCS} and c_2^{DVCS} of Eq. (2.31) would only cancel for a flat acceptance. However, the coefficients are strongly suppressed. The influence of a different acceptance parametrisation is tested in Sect. 7.6.4.

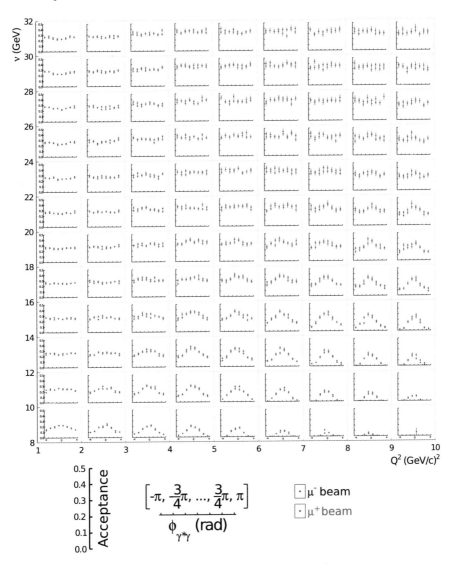

Fig. 7.7 Acceptance for the DVCS process, shown as a function of Q^2, ν and $\phi_{\gamma^*\gamma}$: Each plot in a bin of Q^2 and ν shows the acceptance on the ordinate in eight equidistant bins of $\phi_{\gamma^*\gamma}$ on the abscissa

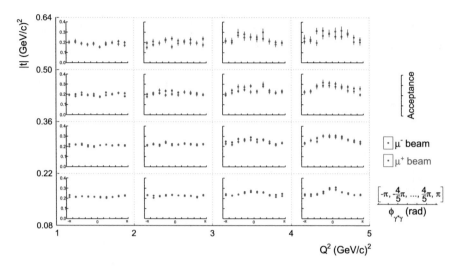

Fig. 7.8 Acceptance for the DVCS process, shown as a function of Q^2, $|t|$ and $\phi_{\gamma^*\gamma}$: Each plot in a bin of Q^2 and $|t|$ shows the acceptance on the ordinate in 10 equidistant bins of $\phi_{\gamma^*\gamma}$ on the abscissa

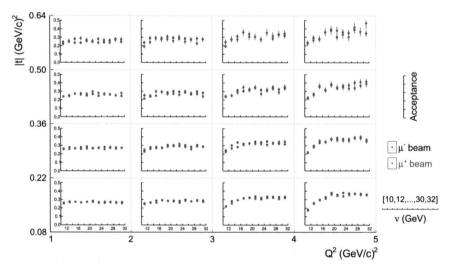

Fig. 7.9 Acceptance for the DVCS process, shown as a function of Q^2, $|t|$ and ν: Each plot in a bin of Q^2 and $|t|$ shows the acceptance on the ordinate in 11 equidistant bins of ν on the abscissa

7.5 The DVCS Cross Section and the Extraction of the *t*-Slope

The DVCS cross section is shown in Fig. 7.10. It is extracted according to Sect. 7.1. The values[9] of the cross section in the four bins of $|t|$ and the corresponding mean kinematic quantities are presented in Table A.2.

To extract the parameter of the *t*-slope, a binned maximum likelihood fit has been used with the following log-likelihood function:

$$\log L(B) = \sum_{n=1}^{4} \sigma_n \log \nu_n(B),$$ (7.11)

where ν_n is given by:

$$\nu_n(B) = \sigma_{tot} \int_{t_n^{min}}^{t_n^{max}} \frac{1}{N} e^{-B|t|} dt.$$

Here t_n^{min} and t_n^{max} denote the edges of the four bins in $|t|$, σ_n the measured cross section in a certain bin n:

$$\sigma_n = \left\langle \frac{d\sigma_{DVCS}^{\gamma^* p \to \gamma p'}}{d|t|} \right\rangle_n \cdot \Delta t_n \quad \text{(see Eq. 6.6),}$$ (7.12)

and $\sigma_{tot} = \sum_{i=n}^{4} \sigma_n$ the total measured cross section. The normalisation N is given by:

$$N = \int_{0.08\,(\text{GeV/c})^2}^{0.64\,(\text{GeV/c})^2} e^{-B|t|} dt.$$

Fig. 7.10 Virtual-photon proton DVCS cross section in the four bins of |t|. Only the statistical errors of the cross section values and the *t*-slope *B* are shown within this plot. After a discussion of systematic uncertainties the final result is presented in Fig. 7.25

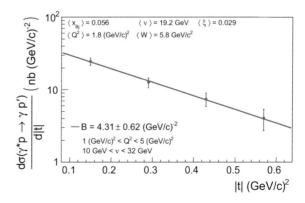

⟨ x_{Bj} ⟩ = 0.056   ⟨ ν ⟩ = 19.2 GeV   ⟨ ξ ⟩ = 0.029
⟨ Q^2 ⟩ = 1.8 (GeV/c)²   ⟨ W ⟩ = 5.8 GeV/c²

— B = 4.31 ± 0.62 (GeV/c)⁻²
1 (GeV/c)² < Q² < 5 (GeV/c)²
10 GeV < ν < 32 GeV

$\frac{d\sigma(\gamma^* p \to \gamma p')}{d|t|}$ (nb (GeV/c)⁻²)

|t| (GeV/c)²

---

[9]The values of the cross section have been cross checked in Ref. [4]

Since one is assuming multinomial statistics within Eq. (7.11), but is dealing with a sum of weights inside each bin instead of a raw number of events, the following error correction has to be applied in order to get correct results for the statistical uncertainty on the t-slope:

$$V_B = (V_{\sum w}) \frac{1}{(V_{\sum w^2})} (V_{\sum w}),$$

The quantity V_B denotes the final variance on the t-slope, $(V_{\sum w})$ the variance given by minimising the log-likelihood using Eq. (7.12) and $(V_{\sum w^2})$ the variance under the exchange:

$$\sigma_n \rightarrow \left(\sum w^2 \right)_n,$$

while $\left(\sum w^2 \right)_n$ has to be calculated according to Eq. (7.6) as follows:

$$\left(\sum w^2 \right)_n = \frac{1}{4} \left[\left(\sum w^2 \right)_n^+ + \left(\sum w^2 \right)_n^- \right],$$

where:

$$
\begin{aligned}
\left(\sum w^2 \right)_n^\pm &= \frac{1}{(\mathcal{L}^\pm \Delta t_n \Delta Q^2 \Delta \nu)^2} \sum_{ij} \left[a_{ijn}^{-2} \left(\sum_{e=1}^{N_{ijn}^{\text{data},\pm}} \frac{1}{\Gamma(Q_e^2, \nu_e)^2} \right. \right. \\
&\quad \left. \left. + (c_{\text{BH}})^2 \cdot \sum_{e=1}^{N_{ijn}^{\text{BH},\pm}} \frac{(w_{\text{P.A.M.}})_e^2}{\Gamma(Q_e^2, \nu_e)^2} + (c_{\pi^0})^2 \cdot \sum_{e=1}^{N_{ijn}^{\pi^0,\pm}} \frac{(w_{\pi^0_\gamma})_e^2}{\Gamma(Q_e^2, \nu_e)^2} \right) \right],
\end{aligned}
\tag{7.13}
$$

with

$$n \in \{1, 2, 3, 4\},$$

representing the four bins in t. In Appendix A.5.6.1 a toy Monte Carlo study is presented, which illustrates the quality of the estimator for the t-slope and the necessity of the error correction in case of weighted events. Furthermore, it shows that the calculated statistical errors are very reasonable. The almost perfect agreement between the data and the exponential fit in Fig. 7.10 may look striking. The p-value, which is in this case the probability to get a better agreement between the data and the model than the present one, is 7%. Appendix A.5.6.1 also shows the χ^2 distribution for a toy MC, which illustrates that the p-value given is correctly calculated. Furthermore, in Appendix A.5.6.2, Fig. 7.10 is shown separately for the two beam charges and for a ν-range from 10 to 20 GeV and 20 to 32 GeV, where one can see statistical fluctuations on a reasonable scale.

7.6 Systematic Uncertainties

Within this section the influence of several systematic effects on the extracted cross section and the t-slope parameter B are studied. The section concludes with a summary and comparison of the different systematic uncertainties.

7.6.1 Variation of the Absolute Normalisation Scale

As shown in Sect. 7.3, a loss of approximately 20% of events for the μ^+ data yield in the region of (80 GeV $< \nu <$ 144 GeV) is observed, when comparing to the pure contribution of the Bethe–Heitler process. Thus, a conservative approach is chosen and it is considered that this loss of events may also be present in the region of (10 GeV $< \nu <$ 32 GeV). Figure 7.11 shows the influence on the extracted value of the t-slope, when one scales the number of measured μ^+ events. It should be emphasised, that this scaling also influences the amount of visible leaking π^0 and thus the amount of π^0 background. However, the amount of the subtracted Bethe–Heitler contribution stays unchanged. In this way it is somehow equivalent to changing the amount of Bethe–Heitler relative to the data and the estimated π^0 contribution. From this a systematic effect on the t-slope $s^{+\downarrow}$:

$$s^{+\downarrow} = 2\%,$$

which preferably lowers the extracted value, is concluded. Figure 7.12 shows the influence of the loss of events in case of the μ^+ data on the extracted cross section in the four bins of t. The systematic effects on the extracted cross section s_i^+ are summarised in Table 7.4.

Furthermore, in Sect. 7.3, in case of the μ^- data yield, an excess of events compared to the Monte Carlo prediction in the region of (80 GeV $< \nu <$ 144 GeV) is detected.

Fig. 7.11 Influence on the result of the extracted value for the t-slope when one rescales the amount of measured events for the μ^+ data. B_0 denotes the preferred value of the t-slope. The plot is normalised to this value. The green band shows the relative statistical uncertainty on the extracted value for the respective scenario

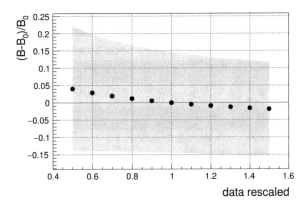

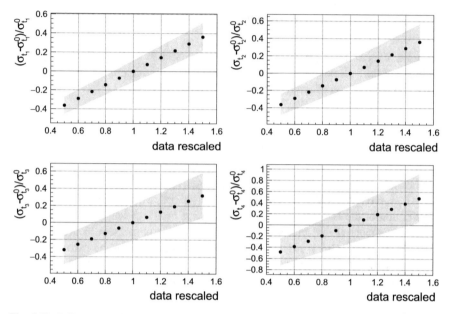

Fig. 7.12 Influence on the result of the measured cross section in the four bins of t when one rescales the amount of measured events for the μ^+ data. $\sigma_{t_i}^0$ denotes the preferred value of the extracted cross section in the corresponding t-bin with $i \in \{1, 2, 3, 4\}$. Each plot is normalised to this value. The green band shows the relative statistical uncertainty on the extracted value for the respective scenario

Table 7.4 Summary of the estimated systematic uncertainties on the cross section in the four bins of t, originating from the uncertainty on the number of measured events in case of the μ^+ data yield. The effect is considered to preferably cause a higher value of the extracted cross section

Bin	σ_{t_1}	σ_{t_2}	σ_{t_3}	σ_{t_4}
Relative sys. error $s_i^{+\uparrow}$	13%	15%	13%	19%

Since the source of this discrepancy is not yet understood, the associated systematic uncertainties are estimated by two different methods:

1. Varying the overall Monte Carlo normalisation.
2. Scaling the μ^- data sample.

For the first method, the influence is comparable to down-scaling the flux since the Monte Carlo is used to calculate the acceptance. This approach leaves the relative amount of the DVCS contribution with respect to the Bethe–Heitler contribution unchanged and simply describes an overall scaling of the extracted cross section. Figure 7.13 shows that the effect is negligible in case of the t-slope extraction. For the cross section in the four bins of t a variation at the order of 10% is considered. Figure 7.14 shows the influence on the cross section when one rescales the flux for the μ^- data. Thus, the following systematic effect on the cross section in the four

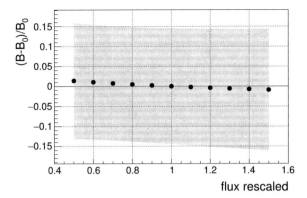

Fig. 7.13 Influence on the result of the extracted value for the t-slope when one rescales the integrated beam flux for the μ^- data. B_0 denotes the preferred value of the t-slope. The plot is normalised to this value. The green band shows the relative statistical uncertainty on the extracted value for the respective scenario

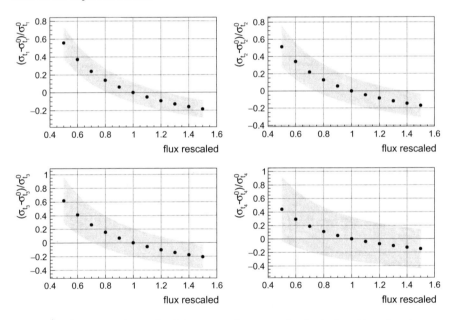

Fig. 7.14 Influence on the result of the measured cross section in the four bins of t when one rescales the integrated beam flux for the μ^- data. $\sigma_{t_i}^0$ denotes the preferred value of the extracted cross section in the corresponding t-bin with $i \in \{1, 2, 3, 4\}$. Each plot is normalised to this value. The green band shows the relative statistical uncertainty on the extracted value for the respective scenario

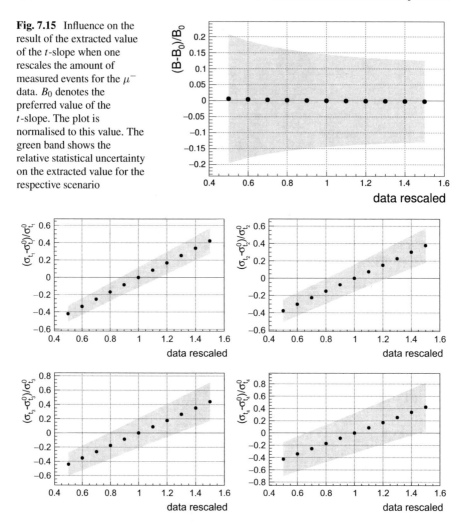

Fig. 7.15 Influence on the result of the extracted value of the t-slope when one rescales the amount of measured events for the μ^- data. B_0 denotes the preferred value of the t-slope. The plot is normalised to this value. The green band shows the relative statistical uncertainty on the extracted value for the respective scenario

Fig. 7.16 Influence on the result of the measured cross section in the four bins of t when one rescales the amount of measured events for the μ^- data. $\sigma^0_{t_i}$ denotes the preferred value of the extracted cross section in the corresponding t-bin with $i \in \{1, 2, 3, 4\}$. Each plot is normalised to this value. The green band shows the relative statistical uncertainty on the extracted value for the respective scenario

bins is concluded:

$$s_i^{-\downarrow} = 6\%, \ i \in \{1, 2, 3, 4\},$$

which preferably lowers the extracted cross section.

The second method to treat the excess of the data observed in case of the μ^- yield corresponds to a scaling of the μ^- data sample. A background effect in the data is assumed and in this case one has to down-scale the number of measured μ^- events, as it was done in the opposite direction for the μ^+ scenario. Figures 7.15 and 7.16 show the influence on the t-slope and the extracted values of the cross section. For the t-slope the effect is again negligible. Considering a down-scaling of 10%, the following systematic uncertainties are concluded for the four values of the cross section:

$$ s_i^{-\downarrow} = 9\%, \ i \in \{1, 2, 3, 4\}, $$

which preferably lower the extracted values. It is not surprising that within the second scenario the systematic uncertainty is larger, compared to the first scenario, since a scaling of the events leaves the estimate of the Bethe–Heitler contribution unchanged. Hence the Bethe–Heitler process has a larger relative impact during the subtraction of its contribution.

The first hypothesis seems to be more plausible. But as one can not be convinced about that, the worst case is absorbed into the systematic uncertainty, which is given by the second scenario. Thus, a nine percent systematic effect on the cross section in the downward direction is assumed.

7.6.2 The π^0 Background Subtraction

In Sect. 7.2 the amount of background, originating from the production of π^0 within the single photon sample, is estimated. This background has two contributions: exclusively produced π^0, described by HEPGen++ and semi-inclusively produced π^0, described by LEPTO. This section evaluates the influence on the results for a variation of the overall π^0 contribution and the fraction of the two production mechanisms within.

Looking at Fig. 7.17, one can see that the systematic effect on the t-slope, originating from the normalisation between the LEPTO and the HEPGen++ Monte Carlo is negligible.

Figure 7.18 shows the influence on the extracted values of the cross section in the four bins of t, originating from the normalisation between LEPTO and HEPGen++. Regarding the summarised results shown in Fig. 7.5, the uncertainty on the contribution of HEPGen++ is considered to be at the order of 20%. The estimates for the systematic error on the cross section in the four bins of t are summarised in Table 7.5.

One of the main systematic uncertainties on the extracted value of the t-slope originates from the absolute normalisation of the total amount of the π^0 background. As described in Sect. 7.2.1 the two Monte Carlo yields describing the π^0 background, are normalised in the first place to the number of visible leaking π^0 in the data. In this case, as described in Sect. 6.2.2, a low energy threshold for the low energetic photons is applied. Figure 7.19 shows the ratio of the number of visible leaking π^0 between data and Monte Carlo as a function of the threshold of the low energy photon after this first normalisation step. From this figure it seems that the Monte Carlo overestimates the π^0 background by up to 30%. Thus, in Figs. 7.20 and 7.21 the influence of a

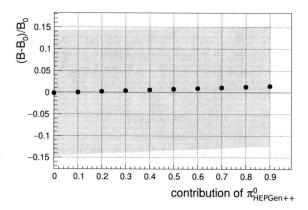

Fig. 7.17 Influence on the extracted value of the *t*-slope, caused by the ratio between the two Monte Carlos describing the π^0 background. The favoured numbers are taken from Sect. 7.2 and give a contribution from HEPGen++ of 10% and a contribution from LEPTO of 90%. The contribution from HEPGen++, $c_{HEPGen++}$, and LEPTO, c_{LEPTO}, is then varied such that $c_{HEPGen++} + c_{LEPTO} = 1$. B_0 denotes the preferred value of the *t*-slope. The plot is normalised to this value. The green band shows the relative statistical uncertainty on the extracted value for the respective scenario

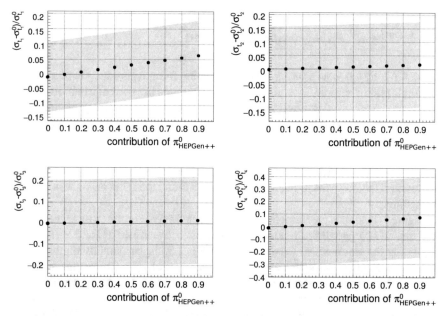

Fig. 7.18 Influence on the extracted value of the cross section in the four bins of *t*, caused by the ratio between the two Monte Carlos describing the π^0 background. The favoured numbers are taken from Sect. 7.2 and give a contribution from HEPGen++ of 10% and a contribution from LEPTO of 90%. The contribution from HEPGen++, $c_{HEPGen++}$, and LEPTO, c_{LEPTO}, is then varied such that $c_{HEPGen++} + c_{LEPTO} = 1$. $\sigma_{t_i}^0$ denotes the preferred value of the extracted cross section in the corresponding *t*-bin with $i \in \{1, 2, 3, 4\}$. Each of the plots is normalised to this value. The green band shows the relative statistical uncertainty on the extracted value for the respective scenario

Table 7.5 Summary of the estimated systematic uncertainties on the cross section in the four bins of t, originating from the normalisation between the LEPTO and the HEPGen++ Monte Carlo yields, used to describe the π^0 background. The effect is considered to preferably cause a higher value of the extracted cross section

Bin	σ_{t_1}	σ_{t_2}	σ_{t_3}	σ_{t_4}
Relative sys. error $s_i^{L,H\uparrow}$	2%	0%	0%	1%

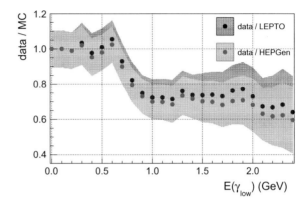

Fig. 7.19 Ratio for the number of visible leaking π^0 events between data and Monte Carlo as a function of the threshold of the low energy photon

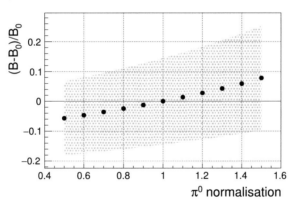

Fig. 7.20 Influence on the extracted value of the t-slope, caused by the normalisation of the π^0 background. B_0 denotes the preferred value of the t-slope. The plot is normalised to this value. The green band shows the relative statistical uncertainty on the extracted value for the respective scenario

possible overestimation of the π^0 background by the Monte Carlo on the t-slope and the extracted values of the cross section is studied. Furthermore, since one observes 116 visible leaking π^0 in the data, and since this number gives the normalisation, the overall statistical uncertainty on the π^0 normalisation is approximately nine percent and has an influence on the extracted values in both directions.

Considering Fig. 7.20 one comes to the following estimates of the systematic uncertainties for the t-slope $s^{\pi^0\downarrow}$ and s^{π^0}:

$$s^{\pi^0\downarrow} = 5\%,$$

and

$$s^{\pi^0} = 2\%,$$

while $s^{\pi^0\downarrow}$ denotes the relative systematic uncertainty due to the definition of the threshold for the low energy photon and s^{π^0} the relative systematic uncertainty due to the statistical uncertainty on the number of visible leaking π^0. The uncertainties on the cross section for the four bins in t, based on Fig. 7.21, are summarised in Tables 7.6 and 7.7.

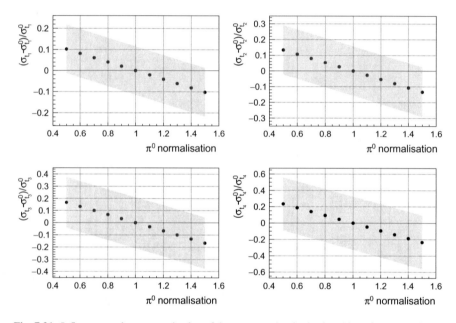

Fig. 7.21 Influence on the extracted value of the cross section in the four bins of t, caused by the normalisation of the π^0 background. $\sigma_{t_i}^0$ denotes the preferred value of the extracted cross section in the corresponding t-bin with $i \in \{1, 2, 3, 4\}$. Each plot is normalised to this value. The green band shows the relative statistical uncertainty on the extracted value for the respective scenario

Table 7.6 Summary of the estimated systematic uncertainties on the cross section in the four bins of t, originating from the uncertainty on the normalisation of the π^0 background, related to the uncertainty on the threshold for the low energy photon. The effect is considered to cause preferably a higher value of the extracted cross section

Bin	σ_{t_1}	σ_{t_2}	σ_{t_3}	σ_{t_4}
Relative sys. error $s_i^{\pi^0\uparrow}$	6%	8%	10%	12%

Table 7.7 Summary of the estimated systematic uncertainties on the cross section in the four bins of t, originating from the statistical uncertainty on the normalisation of the π^0 background

Bin	σ_{t_1}	σ_{t_2}	σ_{t_3}	σ_{t_4}
Relative sys. error $s_i^{\pi^0}$	2%	3%	4%	5%

7.6.3   Radiative Correction Effects

Since radiative corrections are small for the measurement of exclusive single photon production, they are taken into account in the systematic error. A calculation of radiative corrections for COMPASS kinematics of P. A. M. Guichon is used [5]. It provides a reduction of the cross section by factors slightly varying with t, which are reported in Table 7.8. The calculation is done in the one-photon-exchange approximation. However, it is known that the two-photon-exchange will give an opposite effect for μ^+ and μ^-, which cancels for the sum of the two contributions. The application of the factors reported in Table 7.8 to the data provides a slight reduction of the t-slope, as it can be seen in Fig. 7.22.

Table 7.8 Summary of the estimated systematic uncertainties on the cross section in the four bins of t, originating from one-photon-exchange radiative corrections [5]

Bin	σ_{t_1}	σ_{t_2}	σ_{t_3}	σ_{t_4}
Relative sys. error $s_i^{r,\downarrow}$	5.8%	4.7%	4.1%	3.6%

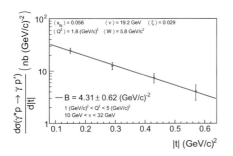 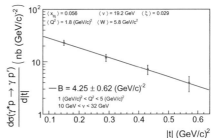

Fig. 7.22 Virtual photon proton DVCS cross section in the four bins of $|t|$, without any influence of radiative corrections (left) and using the estimates of radiative corrections shown in Table 7.8 (right)

7.6.4 Further Scenarios

In this section the influence on the results for different variations of the extraction method is studied. Figure 7.23 shows the influence of different scenarios on the extracted value of the t-slope, while Fig. 7.24 shows the influence on the extracted values of the cross section. The scenarios for which no kinematic fit is used surely should not contribute to the systematic error. In case of the scenarios for which the acceptance binning is changed to a four dimensional binning, including equidistant bins in $\phi_{\gamma*\gamma}$, variations of up to four percent can be observed. Since these variations are not consistent between the scenarios of four and five bins in $\phi_{\gamma*\gamma}$, it is assumed that the effect is rather originating from the fact that the statistics of the Monte Carlo sample, used to calculate the acceptance correction factors, are getting sparse and introduce fluctuations. Thus, it is also not included into the final systematic error.

However, the difference between the kinematic fit with strict energy and momentum balance, denoted by "1phi_fit_!shift", and a shifted energy and momentum balance, denoted by "1phi_fit_shift", should be absorbed into the systematic error since the origin of the shift is unclear (see Sect. 6.3). For the t-slope this is fortunately negligible. In case of the cross section in the four bins of t, the following estimates

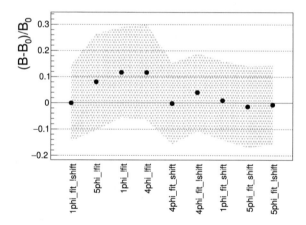

Fig. 7.23 Influence on the extraction of the t-slope for different scenarios: The abbreviation "nphi" denotes the number n of equidistant $\phi_{\gamma*\gamma}$ bins used to extend the acceptance correction factors shown in Fig. 7.9 to a four dimensional acceptance. The fact whether the kinematic quantities corrected by the kinematic fit have been used for the cross section extraction or not is denoted by "fit" and respectively "!fit". In case the kinematic quantities corrected by the kinematic fit have been used for the extraction, it is distinguished, whether the routine is constraint to a strict energy and momentum balance, which is depicted by "!shift" or to a shifted energy and momentum balance, which is depicted by "shift" (see Sect. 6.3). In case of the scenarios "fit", "!fit", "shift" and "!shift" the kinematic quantities Q^2, ν, $\phi_{\gamma*\gamma}$ and t of each event of the final sample have slightly changed and thus the acceptance is individually recalculated for each scenario from a Monte Carlo sample, which has been analysed under the appropriate conditions of the respective scenario. B_0 denotes the preferred value of the t-slope. The plot is normalised to this value. The green band shows the relative statistical uncertainty on the extracted value for the respective scenario

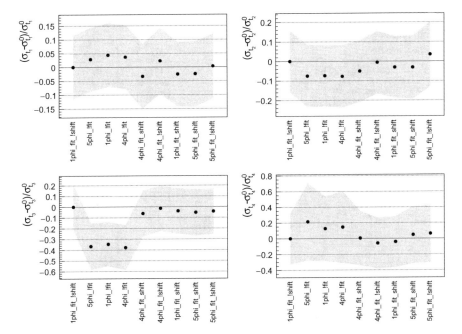

Fig. 7.24 Influence on the extraction of the cross section in the four bins of t for different scenarios. $\sigma_{t_i}^0$ denotes the preferred value of the extracted cross section in the corresponding t-bin with $i \in \{1, 2, 3, 4\}$. Each plot is normalised to this value. The green band shows the relative statistical uncertainty on the extracted value for the respective scenario. For the explanation of the ordinate see Fig. 7.23

of the systematic errors are concluded:

$$s_i^K = 3\%, \ i \in \{1, 2, 3, 4\}.$$

7.6.5   Summary of Systematic Effects

The systematic effects on the cross section in the four bins of t are summarised in Tables 7.9 and 7.10. They are added in quadrature to estimate the final systematic error. The systematic effects on the t-slope are summarised in Tables 7.11 and 7.12. They are added in quadrature to estimate the final systematic error.

This results in the following error bars shown in Fig. 7.25. The inner error bar illustrates the statistical uncertainty, while the outer error bar shows the quadratic sum of the statistical and the systematic uncertainty.

Table 7.9 Summary of the systematic uncertainties on the extracted values of the cross section in the four bins of t, which cause an upward uncertainty

Sections	Effect		σ_{t_1}	σ_{t_2}	σ_{t_3}	σ_{t_4}
7.6.1	Event loss for μ^+ data	$s_i^{+\uparrow} =$	13%	15%	13%	19%
7.6.2	Norm. of π^0 LEPTO/HEPGen++	$s_i^{L,H\uparrow} =$	2%	0%	0%	1%
7.6.2	Threshold uncertainty on π^0	$s_i^{\pi^0\uparrow} =$	6%	8%	10%	12%
7.6.2	Statistical uncertainty on π^0	$s_i^{\pi^0} =$	2%	3%	4%	5%
7.6.4	Muon kinematic uncertainty	$s_i^{K} =$	3%	3%	3%	3%
5.2	Uncertainty on the flux det.	$s_i^{f} =$	3%	3%	3%	3%
	$\sum$	$s_i^{\uparrow} =$	15%	18%	17%	23%

Table 7.10 Summary of the systematic uncertainties on the extracted values of the cross section in the four bins of t, which cause a downward uncertainty

Sections	Effect		σ_{t_1}	σ_{t_2}	σ_{t_3}	σ_{t_4}
7.6.1	MC event loss for μ^- data	$s_i^{-\downarrow} =$	9%	9%	9%	9%
7.6.3	Radiative corrections estimate	$s_i^{r\downarrow} =$	6%	5%	4%	4%
7.6.2	Statistical uncertainty on π^0	$s_i^{\pi^0} =$	2%	3%	4%	5%
7.6.4	Muon kinematic uncertainty	$s_i^{K} =$	3%	3%	3%	3%
5.2	Uncertainty on the flux det.	$s_i^{f} =$	3%	3%	3%	3%
	$\sum$	$s_i^{\downarrow} =$	12%	12%	11%	12%

Table 7.11 Summary of the systematic uncertainties on the extracted value of the t-slope, which cause a downward uncertainty

Sections	Effect		B
7.6.1	Event loss for μ^+ data	$s^{+\downarrow} =$	2%
7.6.2	Norm. of π^0 LEPTO/HEPGen++	$s^{L,H\downarrow} =$	0%
7.6.2	Threshold uncertainty on π^0	$s^{\pi^0\downarrow} =$	5%
7.6.3	Radiative corrections estimate	$s^{r\downarrow} =$	1%
7.6.2	Statistical uncertainty on π^0	$s^{\pi^0} =$	2%
7.6.4	Muon kinematic uncertainty	$s^K =$	0%
	$\sum$	$s^{\downarrow} =$	6%

Table 7.12 Summary of the systematic uncertainties on the extracted value of the t-slope

Sections	Effect		B
7.6.2	Statistical uncertainty on π^0	$s^{\pi^0} =$	2%
7.6.4	Muon kinematic uncertainty	$s^K =$	0%
	$\sum$	$s =$	2%

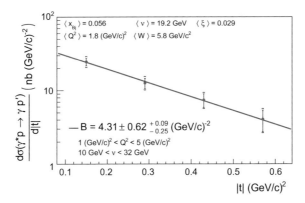

Fig. 7.25 Virtual photon proton cross section in the four bins of t. An exponential fit has been applied from which the t-slope parameter is extracted. The p-value of the exponential fit is 7% and yields in this case the probability to get better agreement than the observed one. The inner error bar illustrates the statistical uncertainty, while the outer error bar shows the quadratic sum of the statistical and the systematic uncertainty. **No radiative corrections are applied but an estimate is included into the systematics**

7.7 Interpretation of the Results

The DVCS cross section has been measured as a function of t. An excellent exponential behaviour of the form:

$$\frac{\mathrm{d}\sigma(\gamma^* p \to \gamma p')}{\mathrm{d}|t|} \propto e^{Bt}, \tag{7.14}$$

is observed within Fig. 7.25. The t-slope parameter B is compared to the H1 and ZEUS measurements, mentioned in Sect. 2.4.5, within Fig. 7.26. The measurements are compatible within the statistical and systematic uncertainties. However, the HERA measurement is much more sensitive on the higher order two gluon exchange, shown in Fig. 2.16, and the comparison might not be completely appropriate. The H1 measurement of B as a function of Q^2, shown in the top right of Fig. 2.17, suggests that the parameter B increases with decreasing values of Q^2. Thus, the COMPASS-II measurement of B, performed at a significantly smaller value of Q^2, might indicate a decrease of B with increasing x_{Bj}. This fact would be in accordance with the reasoning of Sect. 2.3.4 and might give an indication that the transverse size of the nucleon decreases with increasing values of x_{Bj} or $\xi \approx \frac{x_{Bj}}{2-x_{Bj}}$ respectively.

To be more precise, according to Sect. 2.4, the beam charge sum is mainly sensitive to the real and imaginary part of the Compton Form Factor $\mathcal{H}$:

$$\frac{\mathrm{d}\sigma(\gamma^* p \to \gamma p')}{\mathrm{d}|t|}(\xi, t) \propto \mathcal{S}_{CS,U}(\xi, t) \propto c_0^{DVCS}(\xi, t) \propto \mathcal{H}_{\mathrm{Re}}^2(\xi, t) + \mathcal{H}_{\mathrm{Im}}^2(\xi, t).$$

Both, the real and the imaginary part of the Compton Form Factor $\mathcal{H}$ receive most of their contribution from the singlet GPD H_+ at $x = \xi$ in the sense of the Eq. (2.28). Within Sect. 2.3.4 the t-slope $B_{\Delta_\perp^2}$ of the GPD H at $x = \xi$ is related to $< r_\perp^2 >$, the transverse size of the transition matrix element with respect to the centre of momentum of the spectators. Using Eq. (2.24), the measured parameter B can be

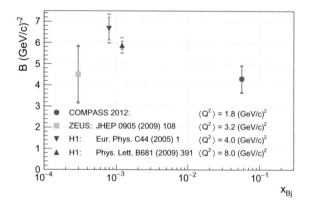

Fig. 7.26 Comparison of the t-slope B, given by Eq. (7.14), extracted by H1 and ZEUS with the result obtained at COMPASS-II. The inner error bars illustrate the statistical uncertainty, while the outer error bars are given by the square root of the quadratic sum of the statistical and systematic uncertainties

COMPASS 2012: $\langle Q^2 \rangle = 1.8$ (GeV/c)^2
ZEUS: JHEP 0905 (2009) 108 $\langle Q^2 \rangle = 3.2$ (GeV/c)^2
H1:   Eur. Phys. C44 (2005) 1 $\langle Q^2 \rangle = 4.0$ (GeV/c)^2
H1:   Phys. Lett. B681 (2009) 391 $\langle Q^2 \rangle = 8.0$ (GeV/c)^2

Fig. 7.27 Comparison of the transverse size of the nucleon $\sqrt{<r_\perp^2>}$, given by Eq. (7.15), extracted by H1 and ZEUS with the result obtained at COMPASS-II. The inner error bars illustrate the statistical uncertainty, while the outer error bars are given by the square root of the quadratic sum of the statistical and systematic uncertainties

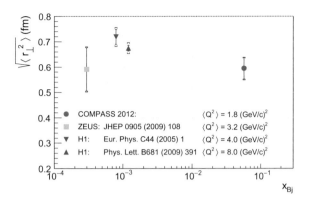

transformed to $<r_\perp^2>$:

$$<r_\perp^2> = 4\hbar^2 B_{\Delta_\perp^2} = 2\hbar^2 \left(\frac{1+\xi}{1-\xi}\right) B = 2\hbar^2 \left(\frac{B}{1-x_{Bj}}\right). \qquad (7.15)$$

It should be noted that the quantity B in Eq. (7.15) is related to B_t from Eq. (2.24) by $B = 2B_t$. This arises from the fact that Eq. (2.24) is valid on the level of amplitudes, while Eq. (7.15) is valid on the level of the cross section.[10] Figure 7.27 shows the conversion of Fig. 7.26 using this reasoning.

In Sect. 2.4.4 the extraction of the proton radius in the valence quark region, according to Ref. [6], is discussed. For this extraction the reasoning is somewhat different. The extracted t-slope values B_t of the imaginary part of the Compton Form Factor $\mathcal{H}$ or respectively the singlet GPD H_+ at $x = \xi$ is mapped to the slope B_0 of the valence GPD H_- at $\xi = 0$ via a correction factor. This correction factor is determined from GPD model studies. The mean valence quark radius squared $<b_\perp^2>$ is given by Eq. (2.34).[11] As the extracted values B_t at $x = \xi$ are extrapolated to $\xi = 0$, the authors of Ref. [6] apply the density interpretation of Sect. 2.3.4, which legitimates the statement of calling $<b_\perp^2>$ a mean proton RMS valence radius. On the left side of Fig. 2.15 the authors compare the extracted radii $<b_\perp^2>$ at different values of x with a Regge inspired ansatz, constraint by the Form Factor F_1 via Eq. (2.20), and state that the data follows the Regge ansatz.

A comparison of the left plot of Fig. 2.15 with the COMPASS-II result on the amplitude level:

$$B_t(\xi = 0.029) = 2.15 \pm 0.31 \, {}^{+0.05}_{-0.13} \, (\text{GeV}/c)^{-2},$$

---

[10]This originates essentially from the following: $(e^{B_t t})^2 = e^{2B_t t} := e^{Bt}$.

[11]To distinguish the amplitude level from the cross section level, the value B_t in this section corresponds to B of Sect. 2.4.4.

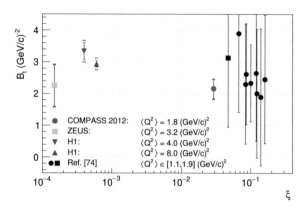

Fig. 7.28 Comparison of the t-slope B_t on the amplitude level extracted by COMPASS-II, H1 and ZEUS with the results of Ref. [6]. The inner error bars of the blue, red and green points illustrate the statistical uncertainty, while the outer error bars are given by the square root of the quadratic sum of the statistical and systematic uncertainties. The black points correspond to the left side of Fig. 2.15. The error bars of the black points mainly originate from the extraction procedure of the imaginary part of the Compton Form Factors $\mathcal{H}$ and not from experimental uncertainties. For a more detailed explanation see Ref. [6]

is shown in Fig. 7.28. The extracted values of B_t are compatible. This is mainly due to the large uncertainties of the black points in Fig. 7.28. It is nevertheless interesting to compare the Regge ansatz, shown within the right plot of Fig. 2.15 with the COMPASS-II result. Several peculiarities arise when one tries to achieve this comparison:

1. To calculate the value $< b_\perp^2 >$ in case of the COMPASS-II data, no correction factor for the extrapolation to $\xi = 0$ is available yet. Thus, inspired by the rather small correction factor of Ref. [6], the ad hoc assumption that the correction factor is small and can be neglected for a first order comparison is made.
2. The COMPASS-II value is in the region of x where the sea quark parton distributions become sizeable with respect to the valence distributions. Thus, in the case of the COMPASS-II data it might be more appropriate to refer to a sea quark radius, given by the singlet GPD H_+.
3. The COMPASS-II result is sensitive to both, the real and the imaginary part of the Compton Form Factor, whereas in case of Ref. [6] a pure contribution from $x = \xi$ is taken into account. As the imaginary part of $\mathcal{H}$ is purely given by the singlet GPD at $x = \xi$, the real part receives contributions from a larger x-region, which is due to the integration within Eq. (2.28). Though the denominator within Eq. (2.26) emphasises the region of $x = \xi$, the real part $\mathcal{H}_{Re}$ can at least in principle even pick up contributions from the ERBL region.
4. The COMPASS-II result is not corrected for radiative effects. However, an estimate is included into the systematic uncertainty on B and the effect of a dedicated treatment of the radiative effects is assumed to be even smaller than the current estimate.

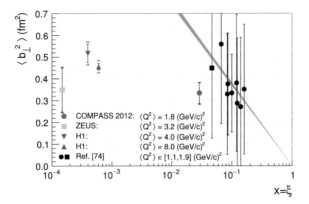

Fig. 7.29 Comparison of the mean transverse nucleon radius squared $< b_\perp^2 >$ of Eq. (7.16) with the Regge Ansatz for valence quarks of Ref. [6]. The red band shows the ansatz of Eqs. (2.35) and (2.36). This ansatz is illustrated together with the black data points by the authors of Ref. [6] in the right plot of Fig. 2.15. The inner error bars of the blue, red and green points illustrate the statistical uncertainty, while the outer error bars are given by the square root of the quadratic sum of the statistical and systematic uncertainties. For the black points only the outer error bars of the right side of Fig. 2.15 are illustrated here

Proceeding nevertheless with the simple ansatz:

$$< b_\perp^2 >= 2B_0\hbar^2 \stackrel{!}{=} 2B\hbar^2, \tag{7.16}$$

the result of Fig. 7.29 is achieved.

With the above remarks in mind, one would be tempted to conclude that the sea quark radius seems to be smaller than the valence quark radius. Regarding additionally the fact that the black data points of Fig. 7.29 do not quantitatively constrain the Regge Ansatz within their large uncertainties, this statement is on a rather weak footing at the moment and future measurements will hopefully give clarification.

Future measurements at JLab 12 GeV will provide more complete information on the eight Compton Form Factors in the valence region. Since the uncertainties extracted in Ref. [6] do not reflect the precision of the measured observables but rather the lack of information on the eight Compton Form Factors, the errors are expected to shrink dramatically.

On the other hand, the 2016/2017 DVCS measurements at COMPASS-II will give complementary information, approaching the valence quark region from the region of the sea quarks. The statistical accuracy in case of the 2016/2017 measurement will increase by approximately a factor of 15 compared to the 2012 pilot run. This will allow to perform the measurement of B for several values of x_{Bj}. Furthermore, a separate measurement of the real and the imaginary part of the Compton Form Factor $\mathcal{H}$ will become feasible as described in Sects. 2.4.2 and 2.4.3 and might give some clarification on the third remark above.

References

1. J. Pretz, Virtual Photon Flux Factor, not neglection the lepton mass (2002). COMPASS Note 2002-11
2. L.N. Hand, Experimental investigation of pion electroproduction. Phys. Rev. **129**, 1834–1846 (1963), https://doi.org/10.1103/PhysRev.129.1834
3. M. Gorzellik, Cross-section measurement of exclusive π^0 muoproduction and firmware design for an FPGA-based detector readout. Dissertation in preparation, Albert Ludwigs Universität Freiburg (2018)
4. A. Ferrero, et al., Extraction of the t-slope of the pure DVCS cross section (2016). COMPASS Note 2016-5
5. P.A.M. Guichon, N. d'Hose, Estimation of the effect of radiative corrections on the measured DVCS cross section, extracted from the 2012 COMPASS-II data (2016). Private communications
6. R. Dupre, M. Guidal, M. Vanderhaeghen, Tomographic image of the proton. Phys. Rev. D **95**, 011501 (2017). https://doi.org/10.1103/PhysRevD.95.011501

Chapter 8
Summary

The structure of the nucleon and in particular its spin decomposition is still puzzling. Though the beautiful concept of Generalised Parton Distributions provides a path to a comprehensive description of the nucleon, these non perturbative, multidimensional objects can only be constrained experimentally. Most of the information on Generalised Parton Distributions is gained within the measurement of Deeply Virtual Compton scattering (DVCS) and Hard Exclusive Meson Production (HEMP). The demanding experimental requirements of high luminosity and a precise detection of all initial and final state particles make the information on Generalised Parton Distributions rather sparse. This opens pioneering ground for experimental physicists.

Within the COMPASS-II programme DVCS and HEMP reactions are currently measured. The most crucial upgrade to the existing COMPASS spectrometer is given by the CAMERA detector, which reconstructs the track of the recoiled target particle and thus ensures the exclusivity of the measurement.

The CAMERA detector was used for the first time as a part of the COMPASS-II spectrometer during a pilot run in 2012. A deep understanding and a precise calibration of the detector prototype is achieved throughout this thesis. The detailed performance studies lead to the exchange of the inner ring of scintillators, which ultimately resulted in an increase of the detector efficiency and a better time resolution. In addition, instabilities within the time synchronisation of the front end electronics with respect to the COMPASS trigger control system and the appearance of random bit-flips within the transmission of the digitised detector signals were identified in the 2012 data. These problems have been tracked down to a particular synchronisation method in the firmware of the GANDALF readout modules. The synchronisation method was successfully reimplemented and the result is a smooth operation of CAMERA since its first commissioning after the 2012 pilot run.

The DVCS analysis of the 2012 data required the most precise and comprehensive determination of the square of the four-momentum transfer to the target proton, which lead to the development of a kinematically constrained fit. This fit makes full use of the exclusive nature of the DVCS and HEMP reactions, providing the most precise determination of the kinematic properties of all involved particles. It is used in many

© Springer International Publishing AG, part of Springer Nature 2018
P. Jörg, *Exploring the Size of the Proton*, Springer Theses,
https://doi.org/10.1007/978-3-319-90290-6_8

ongoing COMPASS-II analysis of exclusive reactions and has become an inevitable part of the longitudinal hit position calibration of the CAMERA detector.

The extraction of the pure DVCS cross section from the data of the 2012 pilot run demands an optimised selection of a single photon sample, the estimation of background contributions originating mainly from $\pi^0 \to \gamma\gamma$, the subtraction of the Bethe–Heitler contribution, the calculation of acceptance correction factors and a careful study of systematic uncertainties. All these steps have been developed and carried out explicitly within this theses.

The final result is shown in Figs. 7.25 and 7.26. It comprises the world's first measurement of the pure DVCS cross section and its exponential dependence B on the square of the four-momentum transfer to the target proton in the region of:

$$1\,(\text{GeV}/c)^2 < Q^2 < 5\,(\text{GeV}/c)^2, \quad 10\,\text{GeV} < \nu < 32\,\text{GeV},$$

$$0.08\,(\text{GeV}/c)^2 < |t| < 0.64\,(\text{GeV}/c)^2,$$

at:

$$< Q^2 > = 1.8\,(\text{GeV}/c)^2, \quad < x_{\text{Bj}} > = 0.056,$$
$$< W > = 5.8\,\text{GeV}/c^2, \quad < \nu > = 19.2\,\text{GeV}.$$

The measurement not only provides valuable data to constrain Generalised Parton Distributions in an uncharted region of x_{Bj}, but one also gains first insight into the evolution of the transverse size of the nucleon on the partonic level, as shown in Fig. 7.27. The extracted t-slope parameter B reads:

$$B_{(x_{\text{Bj}}=0.056)} = 4.31 \pm 0.62 \,^{+0.09}_{-0.25}\,(\text{GeV}/c)^{-2}.$$

It corresponds to a transverse extension $< r_\perp^2 >$ of the nucleon of:

$$\sqrt{< r_\perp^2 >_{(x_{\text{Bj}}=0.056)}} = 2\hbar^2\left(\frac{B}{1 - x_{\text{Bj}}}\right) = 0.60 \pm 0.04 \,^{+0.01}_{-0.02}\,\text{fm}.$$

Further data taking of DVCS reactions through 2016 and 2017 at COMPASS-II will yield approximately a factor of 15 more statistics compared to the 2012 pilot measurement. Hence, the full data set together with the analysis methods developed throughout this thesis will provide the possibility to easily extend the extraction to several values of the parameter B as a function of x_{Bj}. This will reveal the evolution of the transverse size of the nucleon within the COMPASS-II kinematical coverage. In addition, a separate extraction of the real and the imaginary part of the Compton Form Factor $\mathcal{H}$ will become feasible by studying the azimuthal modulations of the cross section.

Chapter 9
Epilog

The detector performance studies in the course of the analysis of the 2012 data lead to vital improvements on the CAMERA prototype in the years following the pilot run. This chapter is supposed to give an overview of the most important improvements applied to the CAMERA detector between 2012 and 2016. The result was the successful detector commissioning during the beginning of the 2016/2017 DVCS data taking, outlined in Appendix A.7, and a smooth operation without any problems since then.

9.1 Replacement of the Inner Scintillators of the CAMERA Detector

As shown in Sect. 5.4, in 2012 a critical compromise had to be made for the high voltage settings of the photomultipliers of the 4 mm thick ring A elements. Setting the high voltage to rather low values causes a decrease of the efficiency at the far end. Since the propagation length of the scintillation light through the scintillator is large, the signals become too small to be detected by the readout electronics. Trying to compensate for the loss on the far end and setting the high voltage to rather large values causes a decrease of the efficiency at the near end, which is due to the fact that the photomultiplier signals become too large and exceed the dynamic range of the readout electronics. This loss in efficiency was the main argument for the exchange of the ring A scintillators. It is a direct consequence of the rather low attenuation lengths of the counters, as it was already detected during their characterisation prior to the 2012 pilot run. The low attenuation lengths are related to defects of the scintillation material, introduced during the manufacturing process. At the time the bad counter performance was detected it was too late though to order a new batch of scintillators for the 2012 pilot run.

© Springer International Publishing AG, part of Springer Nature 2018
P. Jörg, *Exploring the Size of the Proton*, Springer Theses,
https://doi.org/10.1007/978-3-319-90290-6_9

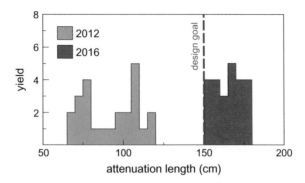

Fig. 9.1 Comparison of the attenuation length of the ring A elements used in 2012 and currently being in use during the 2016/2017 data taking: The ordinate shows the yield of ring A scintillators versus the attenuation length in bins of 10 cm. The values for the attenuation length of the 2012 counters were taken from Ref. [4], while the values for the 2016 counters were taken from Ref. [3]. For the elements used during the 2012 pilot run a large spread of the distribution is visible and during the assembly of ring A in 2012 counters with a rather large attenuation length were placed next to counters showing a rather small value of the attenuation length (see Sect. 5.4)

However, for the 2016 measurement a new batch of scintillators was ordered. Each scintillator was tested individually for its attenuation length[1] during manufacturing and prior to its installation. An element was rejected in case its attenuation length was measured below 150 cm. The measurement of the attenuation length was based on placing a $^{90}_{38}$Sr source[2] on different positions along the scintillators. Analysing the digitised signal amplitude spectra at the different measurement positions allows for an extraction of the attenuation length of a counter. The details of the measurement of the attenuation lengths are given within Refs. [1, 2]. Subsequent to this quality selection a new ring A was assembled and inserted into CAMERA. During the detector commissioning phase the attenuation lengths of each counter were remeasured, using cosmic muons traversing the detector. The results for the attenuation length given by the measurement with the $^{90}_{38}$Sr source and the results obtained with cosmic muons were found to be in excellent agreement [3]. Figure 9.1 shows a comparison of the extracted attenuation lengths for the ring A counters used in 2012 and currently being in use in the 2016/2017 data collecting period. It is clearly visible that the attenuation length has significantly improved, which allowed for a more optimal setting of the high voltage and will finally lead to an overall increase of the efficiency of ring A. Furthermore, the spread of the distribution could be reduced by almost

---

[1]In this context the attenuation length λ is given by the exponential dependence of the photomultiplier amplitudes $A_{up,dwn}$ as a function of the longitudinal hit position z inside a scintillator: $A_{up,dwn} = A_0 \exp(\pm z/\lambda)$, while A_0 is directly proportional to the energy loss of the particle, traversing the scintillator.

[2]Strontium decays into $^{90}_{38}$Y via a β^- decay with a half life of 28.5 y and an average electron energy of 0.196 MeV. $^{90}_{39}$Y decays into $^{90}_{40}$Zr via a β^- decay with a half life of 64.1 h with an average electron energy of 0.934 MeV. The source was chosen because of the rather high average electron energy of the decay into $^{90}_{40}$Zr, necessary to traverse the plastic shielding of the scintillators.

a factor of two, which will lead to more stable values of the efficiency between the different counters.

Though there is no doubt that the efficiency of ring A will increase in 2016, no quantitative results for the increase can be shown here since the analysis of the 2016 data is still in a very early stage.

9.2 Improvements on the CAMERA Readout Electronics

This section will give a short introduction to the readout electronics of the CAMERA detector with the purpose to quickly focus on the improvements to the existing system, developed during this thesis.

9.2.1 Overview

The readout electronics of the CAMERA detector should be capable of instantly extracting signal features, like time-stamp and amplitude information of the analogue photomultiplier pulses, detected at each of the two sides of the 48 scintillators of ring A and B. This has to be achieved for a very wide dynamic range of more than 10 bit for signals from 0 up to -4 V, providing an intrinsic time resolution of the extracted times-stamps at the order of 50 ps.[3] These requirements were the reason for the development of the GANDALF module. Due to its high modularity it is not restricted to the readout of the CAMERA detector, but can also be used as a TDC [5], a scaler [6], a meantimer [7], or a data collector [8]. The basic idea is to divide the module into a powerful mainboard equipped with a Virtex5 SXT FPGA[4] as well as modular mezzanine cards, depending on the specific application. The most complete description of the GANDALF module can be found in Ref. [9].

In case of the CAMERA readout the module is used as a transient recorder and is equipped with two ADC mezzanine cards [10], which provide a 12 bit digitisation of four incoming photomultiplier signals per card with a sampling rate of 933,12 MHz. The digitised data stream is processed on the mainboard and condensed to amplitude and time information [11], which is transmitted to the COMPASS-II data acquisition in case a time correlated trigger signal has been detected. In parallel it is also used for the generation of an independent proton trigger signal [11–13].

Since in total twelve modules are needed for the readout of the detector, the GANDALF modules are operated in a single VXS/VME64x-Crate[5] together with two VXS switch modules, the TIGER modules. The TIGER module consists of

---

[3]This value does not take into account the uncertainty at the order of 300 ps introduced by a photomultiplier together with a ring A or B scintillator. It refers to the achievable resolution of the electronics for an ideal Landau distributed signal generated by e.g. a function generator.

[4]**F**ield-**P**rogrammable **G**ate **A**rray.

[5]VME: **V**ersa **M**odule **E**uropa bus, a computer bus standard; VXS: VME Bus Switched Serial, a computer bus standard which improves the performance of the VME bus.

a custom maid mainboard which comprises amongst other components a Virtex-6 SX315T FPGA, a COM Express CPU, two SFP[6] transceivers, two LEMO outputs and a VXS switch connector. Its FPGA is connected via the VXS backplane to in total 18 payload modules (GANDALF) by $2 \cdot 8$ differential signal pairs per payload module [14].

One of the TIGER modules is used to extract information from the data stream to generate a proton trigger signal, which is crucial for the calibration of the detector. The purpose of the second TIGER module is twofold. First, it transmits the information of the COMPASS trigger control system (TCS [15]) via the VXS backplane to each of the GANDALF modules. Second, it functions as a data concentrator of GANDALF readout.

On each of the GANDALF modules the TCS clock is routed to the main FPGA and both of the two mezzanine cards. Before the 38,88 MHz TCS clock can be used to operate the ADCs on a mezzanine card it is filtered by a clock cleaning and multiplier chip (Si5326 [16]), localised on each of the mezzanine cards. The eight ADCs on one mezzanine card are operated with a 466,56 MHz clock, provided by the Si5326 chip. This results in an effective digitisation of 466,56 GS/s per ADC. To digitise one analogue signal, two ADCs are used in interleaved mode with clocks shifted by π. Thus, a final digitisation of 933,12 GS/s is reached.

Combined with a data ready signal the 12 bit data word from each digitisation step is passed to the GANDALF main FPGA for further processing of the data. This processing has to happen synchronous to a single clock. We have chosen the data ready signal of the first ADC on the upper mezzanine card. In the following, the data ready signal of the upper of the two mezzanine cards will be denoted as SiA data ready signal and respectively in case of the lower mezzanine card as SiB data ready signal.

The crucial ingredients for a correct sampling of the 12 bits per ADC are:

- A fixed phase relation between the output clocks of the Si5326 chips, operating the ADCs on the two mezzanine cards.
- A correct setting of the I/O delays[7] of the FPGA for each of the $16 \cdot 12$ bits per module.

The first point is a necessity for the second one. Otherwise, the I/O delay values of the lower card will change from one initialisation to another.

Different from what has been claimed in Ref. [17] a readout of the ADCs free from errors was not possible. Therefore we had to develop a new method to achieve the time synchronisation of the 16 ADC chips. The details and the new method will be explained in the following section.

---

[6]Small Form-factor Pluggable: A specification of a generation of modular optical or electronical transceivers.

[7]An I/O delay element is a common building block inside an FPGA. It delays the Input/Output signal accessing or leaving the FPGA. It is adjusted to guarantee that the in/output signal is synchronous to the sampling clock.

9.2.2 Time Synchronisation

During the analysis of the 2012 data it turned out to be necessary to calibrate the bias of the time of flight between an inner and outer scintillator for each of the 48 scintillator combinations run by run. This is related to the fact that inner and outer scintillators are read out by different GANDALF modules, which have to be synchronised in a correct way. Furthermore, approximately 10 percent of the data had to be excluded from the analysis, due to biterrors in the sampling process within the data transfer from ADC to FPGA. The fact that the time of flight offset can change after a restart of the FPGAs and the appearance of biterrors in the data transfer are closely connected.

Figure 9.2 defines[8] on the left side an ideal initialisation of the GANDALF FPGA. The TCS clock, provided by the COMPASS trigger control system, is exactly in phase with the two data ready signals of the upper ADC on each of the two mezzanine cards. The time measurement is uniquely synchronised to the rising edge of the TCS clock.

Looking at the right side of Fig. 9.2, a typical initialisation without any synchronisation mechanism is shown. The phase of the two data ready signals of the ADCs with respect to the TCS clock and with respect to each other is arbitrary. The time measurement will thus have an arbitrary offset after each reinitialisation, which is denoted as $\Delta t_{A,B}$. But even more severe is the fact that, depending on the arbitrary phase, it may happen that the bits transmitted from the ADCs to the FPGA are

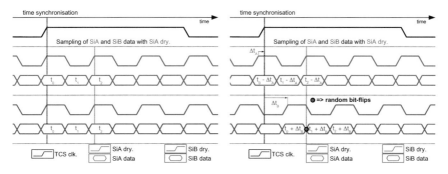

Fig. 9.2 Time synchronisation and sampling of the data transmitted from the ADCs to the GAN-DALF main FPGA. The left side defines an ideal FPGA initialisation, for which the internal operating clock of the FPGA is given by the SiA data ready signal, which is synchronised to the TCS clock transmitted by the COMPASS trigger control system. The data of both ADCs is correctly sampled with the SiA data ready signal. The chronological order of the data is depicted by the time-stamps t_i. The right side shows a typical initialisation without any synchronisation mechanism applied. The time synchronisation shows an offset $\Delta t_A/\Delta t_B$ for both data ready signals and the SiB data is sampled outside its data eye by the SiA data ready

---

[8]It is not necessary that the rising edge of the TCS clock is exactly aligned to the rising edges of the SiA/B data ready signals. This scenario is simply chosen for pedagogical reasons. The crucial point is that the phase is fixed from one initialisation to another and that the I/O delays are correctly chosen for each of the $8 \cdot 12$ bits per module, according to this fixed phase relation.

wrongly sampled. Figure 9.2 shows the corresponding scenario on the right side for the SiB data, which is sampled with the SiA data ready outside its data eye, causing arbitrary bit-flips. The appearance of this scenario is due to the fact that in contrary to the statement in the first data sheet of the chip, from one initialisation to another the Si5326 chip can not provide a fixed phase between its input (TCS) and output (SiA/B) clock. It provides however the possibility to shift the output clock with respect to the input clock in steps of 4.06 ps. This gives rise to the new synchronisation method.

New Synchronisation Method

The steps of the new synchronisation method are shown schematically in Fig. 9.3. A combination of a VHDL[9] module implemented inside the FPGA firmware [18] and computations on the CPU in the VXS/VME64x-Crate is used. The basic idea to achieve the clock synchronisation after a reload of the FPGA firmware is illustrated within the blue box of Fig. 9.3. Two flip-flops[10] inside the FPGA are driven with the TCS clock itself. The flip-flops can be regarded to have a fixed latency with respect

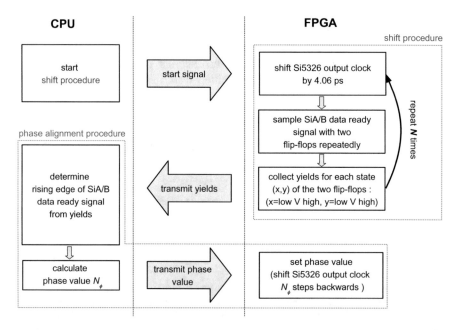

Fig. 9.3 Schematic illustration of the clock synchronisation method of a GANDALF module. The number of shifts steps N within the blue box is set such that more than two clock periods of the SiA/B data ready signals are covered by the procedure (see e.g. top right and left distributions of Fig. 9.4)

---

[9] **V**ery **H**igh **S**peed **I**ntegrated **C**ircuit **H**ardware **D**escription **L**anguage.

[10] A flip-flop is a bistable multivibrator being a fundamental building block of digital electronics. Inside an FPGA it is driven by a clock sampling the state of an input signal. During each cycle of the driving clock the output of the flip-flop yields the current state (low or high) of the input signal.

to the TCS clock and thus with respect to each other. The SiA/B data ready is used as the input signal for the two flip-flops. Shifting the SiA/B clock, the output clock of the Si5326 chip, and as a consequence the SiA/B data ready signals in steps of 4.06 ps, one observes different scenarios for the state of the two flip-flops. For each 4.06 ps step a large amount of statistics is taken and the yields for which the state of the two flip-flops was either (low, low), (high, high), (high, low) or (low, high) are transmitted to the CPU. This marks the beginning of the phase alignment procedure, illustrated by the red box within Fig. 9.3. The analysis of the transmitted yields on the CPU allows for a precise determination of the phase between the TCS clock and the SiA/B data ready signal, as it will be described in detail within the next paragraph. The appropriate number of shift steps, which is given by the value of the phase in units of 4.06 ps, is then transmitted back to the FPGA. Next, the correct phase for the SiA/B clock is set by the FPGA firmware to bring the system to a state as shown on the left side of Fig. 9.2.

Phase Alignment Procedure

Details on the determination of the phase between the SiA/B data ready signals and the TCS clock as well as the accuracy of the method are illustrated in Fig. 9.4. The top left distribution shows the number of occurrences in shaded grey for which both flip-flops detected a high state of the SiA/B data ready signal. The grey distribution and the red fit function show the number of occurrences for which the two flip-flops detected either a (low, high) or a (high, low) state of the SiA/B data ready signal while sampling a rising edge. The top right distribution shows the number of occurrences of (low, high) or (high, low) states, while the first two distributions corresponding to a rising edge are marked by a red fit function. The middle left distribution shows a zoom on the second of these two distributions. It gives a precise characterisation of a rising edge of the SiA/B signal. Since the rectangular shape might not be completely intuitive its emergence is described in Appendix A.6 inside Fig. A.59.

The goal of the procedure is to determine the value of the phase between the SiA/B data ready signal and the TCS clock. As shown within the blue box of Fig. 9.3 in total N shift steps have been performed inside the FPGA. Denoting the location of the rising edge shown in the middle left plot of Fig. 9.4 by S, the value of the phase in units of the 4.06 ps steps is given by[11]:

$$N_\phi = S - N.$$

This corresponds to performing N_ϕ shift steps backwards within the FPGA to achieve the phase alignment. The problem of the determination of the phase value is thus reduced to a determination of a unique time-stamp S of a rising edge of the SiA/B data ready signal, which is explained in the following:

The red fit function, shown within the upper three distributions of Fig. 9.4 has the following form:

$$f(x; N, \mu_1, \sigma_1, \mu_2, \sigma_2) := M\Big(R(x; \mu_1, \sigma_1) + F(x; \mu_2, \sigma_2) + \Theta(x - \mu_1)\Theta(\mu_2 - x)\Big),$$

---

[11]One could have chosen equally well the first or last rising edge within the top right plot of Fig. 9.4, since N_ϕ has to be determined only up to multiples of the SiA/B signal period.

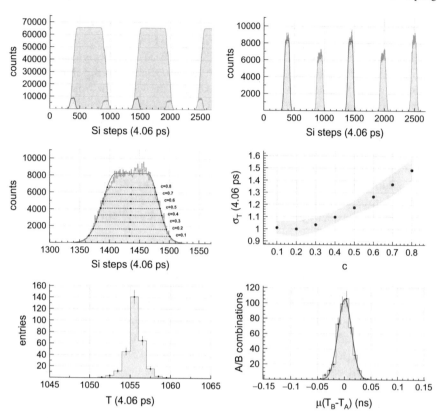

Fig. 9.4 Illustration of the steps performed for the detection of a rising edge of the SiA/B data ready signals: Top left: Occurrences when both flip-flops, sampling the SiA/B data ready signal, have detected a high state (grey shaded), occurrences when the two flip-flops have detected a different state of the SiA/B data ready signal and a rising edge is identified (grey and red fit). Top right: Occurrences when both flip-flops have detected a different state of the SiA/B data ready signal (see Fig. A.59). The rising edges are marked by a red fit function. Middle left: Zoom on one of the rising edges of the top right distribution. The different time-stamps of the distribution for different values of the fraction c are shown by the black dots in the centre of the distribution. The fit function and the fraction c are explained in detail inside the text of Sect. 9.2.2. Middle right: Width of the distributions of the SiA/B data ready signal period for different values of c. Bottom left: Distribution of the SiA/B data ready signal period for $c = 0.2$. Bottom right: Distribution of the mean value of the time of flight spectra between ring A and B of the CAMERA detector. The gaussian fit has a σ of 12.6 ps. Details on this measurement can be found inside the text of Sect. 9.2.2

while the functions R and F are defined as follows:

$$R(x; \mu_1, \sigma_1) := \exp\left(-\left(\frac{x - \mu_1}{\sigma_1}\right)^2\right)\Theta(\mu_1 - x),$$

$$F(x; \mu_2, \sigma_2) := \exp\left(-\left(\frac{x - \mu_2}{\sigma_2}\right)^2\right)\Theta(x - \mu_2).$$

The Heaviside function is denoted Θ, while $\mu_{1,2}$ and $\sigma_{1,2}$ denote the mean values and standard deviations of the two Gaussian functions R and F. The normalisation is given by the parameter M. Instead of simply choosing μ_1 or μ_2 as the time-stamp it has been found that a more precise time-stamp S can be extracted using the following definitions:

$$S := \frac{T_R + T_F}{2},$$

$$T_R := R^{-1}(c; \mu_1, \sigma_1),$$

$$T_F := F^{-1}(c; \mu_2, \sigma_2).$$

The functions R^{-1} and F^{-1} denote the numerically calculated inverse functions of R and F defined on the interval $[-\infty, \mu_1[$ and respectively $]\mu_2, \infty]$.

The role of the position c at which the functions R^{-1} and F^{-1} are evaluated is also shown in the middle left plot of Fig. 9.4. Basically the values T_R and T_F denote the positions on the abscissa for which the value of the function f on its rising or respectively falling edge equals the fraction c of the maximum plateau value M. The time-stamp S for a given value of c is shown by the black point in the middle of the distribution. In order to find the value of c, which yields the most precise time-stamp S, the synchronisation procedure of Fig. 9.3 has been repeated a few 100 times. For each cycle two time-stamps S_1 and S_2, characterising two subsequent rising edges, have been extracted. The distribution of $T = S_2 - S_1$, which characterises the inverse clock frequency of the SiA/B data ready signal in units of the Si5326 shift steps for a c-value of 0.2 is shown at the bottom left of Fig. 9.4. The c-value has been determined by studying the width of the distributions of T for different values of c as shown in the middle right plot of Fig. 9.4. From this point of view the precision of the alignment of a rising edge of the SiA/B data ready signal, $\sigma_1(S)$, is at the order of one Si5326 step, which corresponds to:

$$\sigma_1(S) \approx \frac{4.06\,\text{ps}}{\sqrt{2}} \approx 2.9\,\text{ps}.$$

The factor $\sqrt{2}$ arises from the fact that $\sigma(T) = \sqrt{\sigma(S_2)^2 + \sigma(S_1)^2} = \sqrt{2}\,\sigma(S)$ has been measured.

To get an ultimate confirmation that the whole synchronisation procedure is sufficiently precise and reliable, the final setup used for the readout of the CAMERA detector in 2016 has been tested with a laser system. A laser pulse is injected simultaneously in the middle of all the ring A and ring B scintillators of the CAMERA detector. By measuring the time of flight $(T_B - T_A)$ as defined in Sect. 2.5, the mean values $\mu(T_B - T_A)$ of the time of flight spectra for a sequence of reloads can be extracted. They are shown in the bottom right distribution of Fig. 9.4. The resolution of the time of flight offset $\sigma_{ToF} := \sigma(\mu(T_B - T_A))$ from one initialisation of the system to another is given by:

$$\sigma_{ToF} \approx 12.8\,\text{ps}.$$

Considering the fact that both ends of a scintillator are influenced by a single alignment procedure and that ring A and B elements are subject to different phase alignment procedures, the distribution of the time of flight offset is sensitive to two individual alignment procedures. Thus, σ_{ToF} corresponds to an uncertainty on a single phase alignment procedure of:

$$\sigma_2(S) \approx \frac{\sigma_{ToF}}{\sqrt{2}} \approx 9.0\,\text{ps}.$$

The discrepancy to $\sigma_1(S)$ might be due to systematic effects on the extraction of the mean values of the time of flight spectra, caused by instabilities of the laser system. It has not been further investigated since for a time of flight resolution at the order of 300 ps, an offset from one initialisation procedure to another of 12.8 ps is sufficiently precise. Furthermore, it is worth mentioning that there are no measurements outside the range of the bottom right distribution of Fig. 9.4, which confirms that the procedure has not failed once.

References

1. V. Behrendt, Characterization of the CAMERA Detector at the COMPASS II Experiment. Bachelor thesis, Albert Ludwigs Universität Freiburg (2015)
2. C. Böhm, Characterisation of a Recoil Proton Detector at the COMPASS II Experiment. Bachelor thesis, Albert Ludwigs Universität Freiburg (2015)
3. S. Scherrers, The CAMERA Detector for the COMPASS-II Experiment at CERN - Time and Amplitude Calibrations with Lasers and Muons. Bachelor thesis, Albert Ludwigs Universität Freiburg (2016)
4. R. Schäfer, Charakterisierung eines Detektors zum Nachweis von Rückstoßprotonen am COMPASS Experiment. Diploma thesis, Albert Ludwigs Universität Freiburg (2013)
5. M. Büchele, Entwicklung eines FPGA-basierten 128-Kanal Time-to-Digital Converter für Teilchenphysik-Experimente. Diploma thesis, Albert Ludwigs Universität Freiburg (2012)
6. C. Michalski, Entwicklung eines Echtzeit-Strahlprofill-Monitoring-Systems für das COMPASS-II-Experiment. Diploma thesis, Albert Ludwigs Universität Freiburg (2013)
7. J. Bieling, Entwicklung eines ungetakteten 64-Kanal-Meantimers und einer Koinzidenzschaltung auf einem FPGA. Diploma thesis, Universität Bonn (2010). https://www.rwth-aachen.de/global/show_document.asp?id=aaaaaaaaaamlakm
8. T. Grussenmeyer, Entwicklung eines modularen und verteilten Datenaufnahmesystems für Testexperimente. Diploma thesis, Albert Ludwigs Universität Freiburg (2013)
9. F. Herrmann, Development and Verification of a High Performance Electronic Readout Framework for High Energy Physics. Dissertation, Albert Ludwigs Universität Freiburg (2011). https://freidok.uni-freiburg.de/data/8370, URN: urn:nbn:de:bsz:25-opus-83709
10. S. Schopferer, Entwicklung eines hochauflösenden Transientenrekorders. Diploma thesis, Albert Ludwigs Universität Freiburg (2009)
11. P. Jörg, Untersuchung von Algorithmen zur Charakterisierung von Photomultiplierpulsen in Echtzeit. Diploma thesis, Albert Ludwigs Universität Freiburg (2013)
12. T. Baumann, Entwicklung einer Schnittstelle zur Übertragung von Pulsinformationen von einem Rückstoßdetektor an ein digitales Triggersystem. Diploma thesis, Albert Ludwigs Universität Freiburg (2013)
13. M. Gorzellik, Entwicklung eines digitalen Triggersystems für Rückstoßproton Detektoren. Diploma thesis, Albert Ludwigs Universität Freiburg (2013)

14. S. Schopferer, An FPGA-based Trigger Processor for a Measurement of Deeply Virtual Cómpton Scattering at the COMPASS-II Experiment. Dissertation, Albert Ludwigs Universität Freiburg (2013), https://freidok.uni-freiburg.de/data/9274, URN: urn:nbn:de:bsz:25-opus-92742

15. B. Grube, A Trigger Control System for COMPASS and a Measurement of the Transverse Polarization of Λ and Ξ Hyperons from Quasi-Real Photo-Production. Dissertation, Technische Universität München (2006). https://mediatum.ub.tum.de/doc/603118/603118.pdf

16. Silicon Laboratories, Any-frequency precision clocks Si53xx family reference manual (2010). www.silabs.com

17. P. Kremser, Optimierung und Charakterisierung eines Transientenrekorders für Teilchenphysikexperimente. Diploma thesis, Albert Ludwigs Universität Freiburg (2014)

18. M. Gorzellik, Cross-section measurement of exclusive π^0 muoproduction and firmware design for an FPGA-based detector readout. Dissertation in preparation, Albert Ludwigs Universität Freiburg (2018)

Appendix A

A.1 CAMERA Calibration

A.1.1 Azimuthal Angle Calibration

See Figs. A.1 and A.2.

© Springer International Publishing AG, part of Springer Nature 2018
P. Jörg, *Exploring the Size of the Proton*, Springer Theses,
https://doi.org/10.1007/978-3-319-90290-6

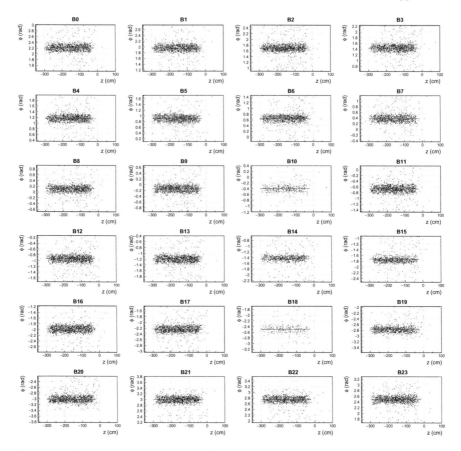

Fig. A.1 Azimuthal angle ϕ as a function of the reconstructed z-position of the hits detected inside the 24 B counters of CAMERA. The counter number is indicated inside the respective distribution

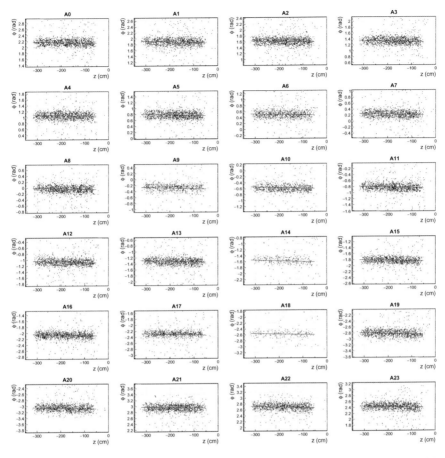

Fig. A.2 Azimuthal angle ϕ as a function of the reconstructed z-position of the hits detected inside the 24 A counters of CAMERA. The counter number is indicated inside the respective distribution

A.1.2   Calibration of the Longitudinal Position

See Figs. A.3 and A.4.

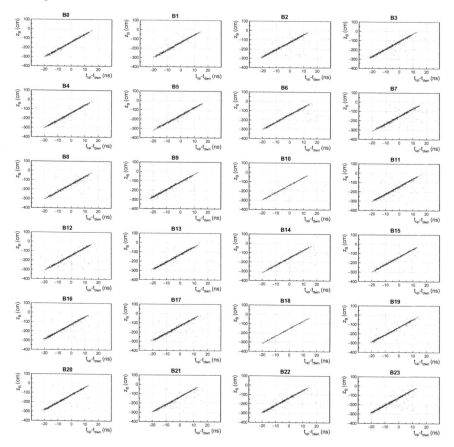

Fig. A.3 Distribution of the predicted z-position z_B of the recoiled particle inside the 24 B counters of CAMERA as a function of $(t_u - t_d)$, the difference of the up- and downstream time-stamps measured with the two photomultiplier tubes of the respective counter. The quantity z_B has been predicted by the usage of the kinematically constrained fit of Sect. 5.1.2. The counter number is indicated inside the respective distribution

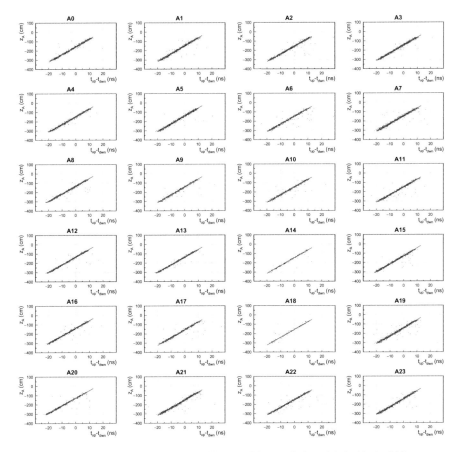

Fig. A.4 Distribution of the predicted z-position z_A of the recoiled particle inside the 24 B counters of CAMERA as a function of $(t_u - t_d)$, the difference of the up- and downstream time-stamps measured with the two photomultiplier tubes of the respective counter. The quantity z_A has been predicted by using an interpolation between the interaction vertex and the hit position in ring B. The counter number is indicated inside the respective distribution

A.2   CAMERA Efficiency

A.2.1   Ring A Efficiencies

Upstream Side

See Figs. A.5, A.6, A.7 and A.8.

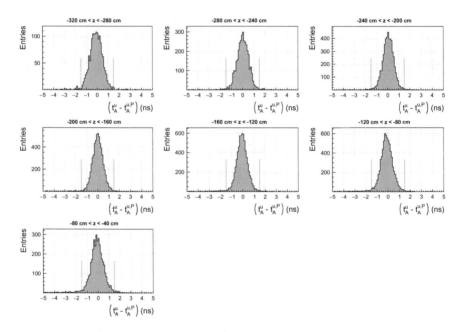

Fig. A.5 Distributions corresponding to Fig. 5.15 for several bins in the longitudinal hit position z. The range in z is indicated within the distributions

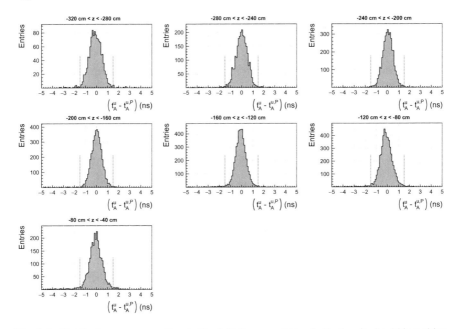

Fig. A.6 Distributions corresponding to Fig. 5.15 for several bins in the longitudinal hit position z. The range in z is indicated within the distributions. Only the data yield taken with the μ^- beam is used

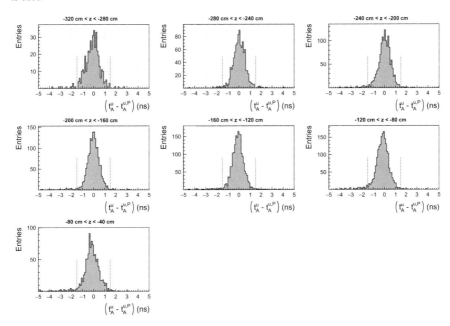

Fig. A.7 Distributions corresponding to Fig. 5.15 for several bins in the longitudinal hit position z. The range in z is indicated within the distributions. Only the data yield taken with the μ^+ beam is used

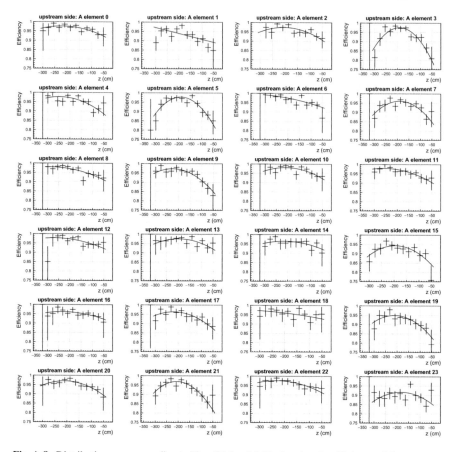

Fig. A.8 Distributions corresponding to Figs. 5.16 and 5.17, showing the efficiency of the upstream side of ring A individually for each scintillator, as indicated inside the distributions. The black curves show the parametrisations used to include the efficiency into the simulations

Downstream Side

See Figs. A.9, A.10, A.11, A.12 and A.13.

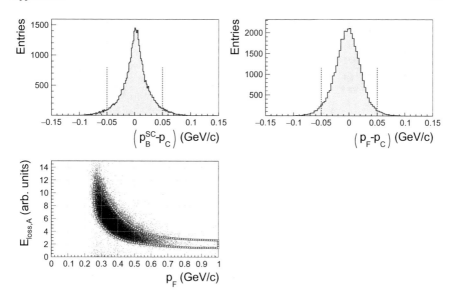

Fig. A.9 Distributions corresponding to Figs. 5.13 and 5.14 for the selection of the N_0 sample described in Sect. 5.4, but in case of the extraction of the downstream efficiency of ring A, using the measured upstream time-stamp

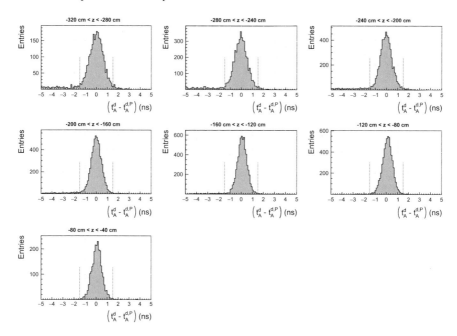

Fig. A.10 Distributions corresponding to Fig. 5.15 for several bins in the longitudinal hit position z and in case of the downstream side efficiency determination of ring A. The range in z is indicated within the distributions

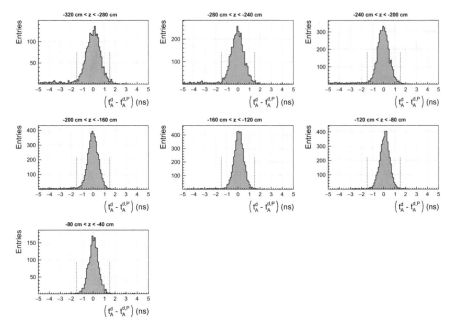

Fig. A.11 Distributions corresponding to Fig. 5.15 for several bins in the longitudinal hit position z and in case of the downstream side efficiency determination of ring A. The range in z is indicated within the distributions. Only the data yield taken with the μ^- beam is used

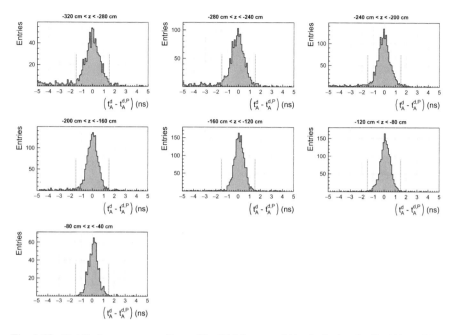

Fig. A.12 Distributions corresponding to Fig. 5.15 for several bins in the longitudinal hit position z and in case of the downstream side efficiency determination of ring A. The range in z is indicated within the distributions. Only the data yield taken with the μ^+ beam is used

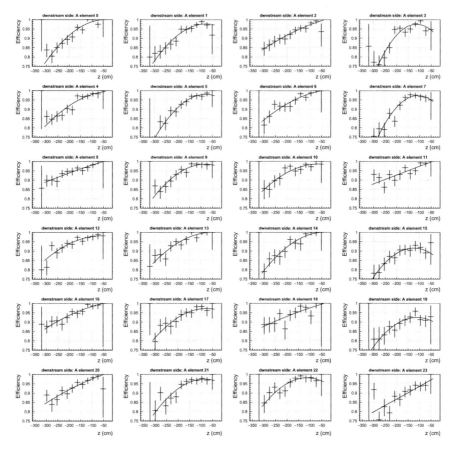

Fig. A.13 Distributions corresponding to Figs. 5.16 and 5.17, showing the efficiency of the downstream side of ring A individually for each scintillator, as indicated inside the distributions. The black curves show the parametrisations used to include the efficiency into the simulations

A.2.2 Ring B Efficiencies

See Fig. A.14.

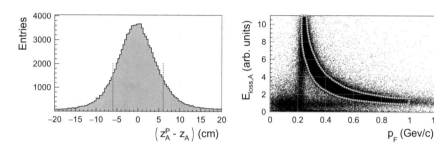

Fig. A.14 Distributions corresponding to Figs. 5.11 and 5.12, in case of the ring B efficiency determination, while no inter-calibration with the startcounter is available in this case. Left: Distribution of the difference between the predicted z-position z_A^P in ring A using the kinematically constrained fit and the reconstructed z-position z_A determined by the up and down time-stamps of ring A. Right: Distribution of the energy loss in ring A, $E_{loss,A} = \sqrt{A_A^u A_A^d}$, as a function of the proton momentum p_F deduced with the kinematically constrained fit. The blue polygon indicates the cut applied in order to select the N_O sample corresponding to Eq. (5.12). The blue lines indicate the cuts applied in order to select the N_O sample corresponding to Eq. (5.12)

Upstream Side

See Figs. A.15, A.16, A.17, A.18 and A.19.

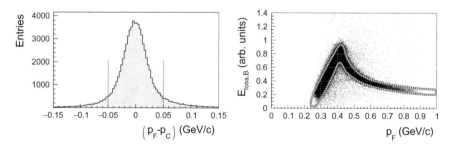

Fig. A.15 Distributions corresponding to Figs. 5.13 and 5.14, in case of the determination of the efficiency of the upstream side of ring B, while no inter-calibration with the startcounter is available in this case. Left: Distribution of the difference between the proton momentum p_F given by the kinematically constrained fit and the proton momentum p_C, defined in Sect. 5.4. Right: Distribution of the energy loss $E_{loss,B} = \sqrt{A_B^{u,P} A_B^d}$ as a function of p_F, the proton momentum deduced with the kinematically constrained fit. The quantity $A_B^{u,P}$ denotes the predicted upstream amplitude in ring B given by Eq. (5.13), while A_B^d denotes the measured downstream amplitude in ring A. The blue lines indicate the cuts applied in order to select the N_O sample corresponding to Eq. (5.12)

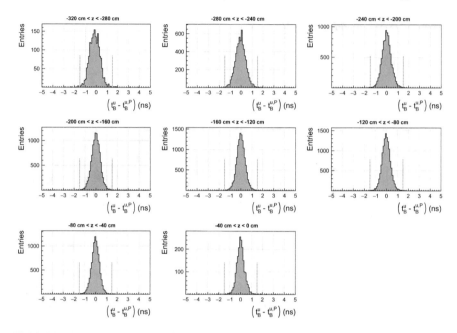

Fig. A.16 Distributions corresponding to Fig. 5.15 for several bins in the longitudinal hit position z, in case of the determination of the efficiency of the upstream side of ring B. The range in z is indicated within the distributions

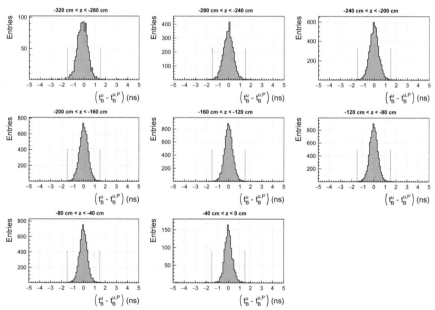

Fig. A.17 Distributions corresponding to Fig. 5.15 for several bins in the longitudinal hit position z, in case of the determination of the efficiency of the upstream side of ring B. The range in z is indicated within the distributions. Only the data yield taken with the μ^- beam is used

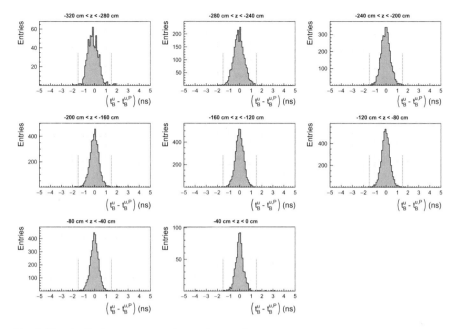

Fig. A.18 Distributions corresponding to Fig. 5.15 for several bins in the longitudinal hit position z, in case of the determination of the efficiency of the upstream side of ring B. The range in z is indicated within the distributions. Only the data yield taken with the μ^+ beam is used

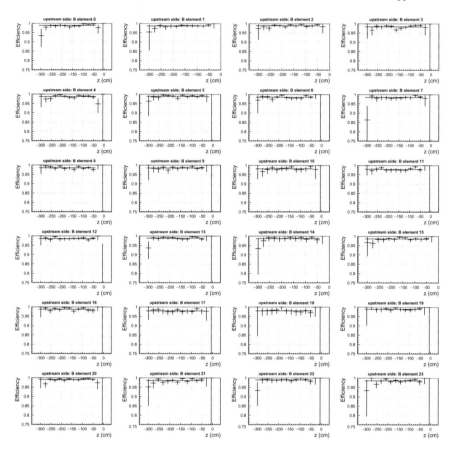

Fig. A.19 Distributions corresponding to Fig. 5.18, showing the efficiency of the upstream side of ring B individually for each scintillator, as indicated inside the distributions. The black curves show the parametrisations used to include the efficiency into the simulations

Downstream Side

See Figs. A.20, A.21, A.22, A.23 and A.24.

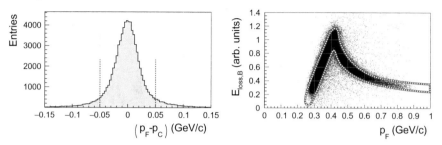

Fig. A.20 Distributions corresponding to Figs. 5.13 and 5.14, in case of the determination of the efficiency of the downstream side of ring B, while no inter-calibration with the startcounter is available in this case. Left: Distribution of the difference between the proton momentum p_F given by the kinematically constrained fit and the proton momentum p_C, defined in Sect. 5.4. Right: Distribution of the energy loss $E_{loss,B} = \sqrt{A_B^{u,P} A_B^d}$ as a function of p_F, the proton momentum deduced with the kinematically constrained fit. The quantity $A_B^{d,P}$ denotes the predicted downstream amplitude in ring B given by Eq. (5.13), while A_B^d denotes the measured downstream amplitude in ring A. The blue lines indicate the cuts applied in order to select the N_O sample corresponding to Eq. (5.12)

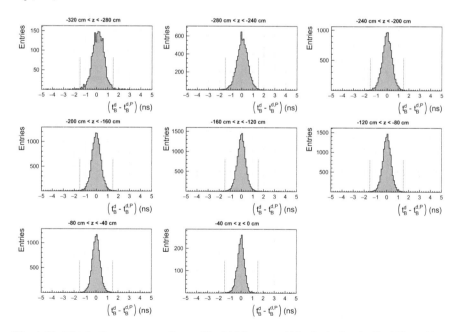

Fig. A.21 Distributions corresponding to Fig. 5.15 for several bins in the longitudinal hit position z, in case of the determination of the efficiency of the downstream side of ring B. The range in z is indicated within the distributions

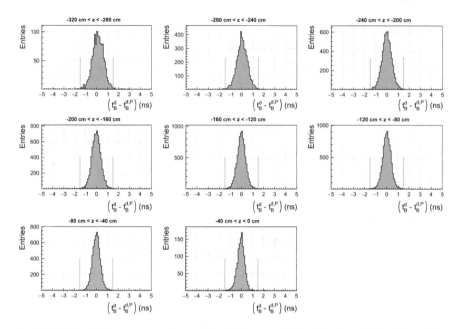

Fig. A.22 Distributions corresponding to Fig. 5.15 for several bins in the longitudinal hit position z, in case of the determination of the efficiency of the downstream side of ring B. The range in z is indicated within the distributions. Only the data yield taken with the μ^- beam is used

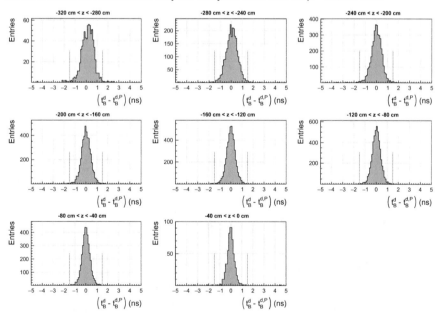

Fig. A.23 Distributions corresponding to Fig. 5.15 for several bins in the longitudinal hit position z, in case of the determination of the efficiency of the downstream side of ring B. The range in z is indicated within the distributions. Only the data yield taken with the μ^+ beam is used

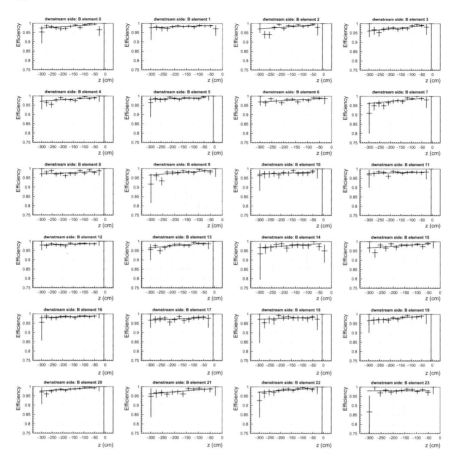

Fig. A.24 Distributions corresponding to Fig. 5.18, showing the efficiency of the downstream side of ring B individually for each scintillator, as indicated inside the distributions. The black curves show the parametrisations used to include the efficiency into the simulations

A.3 Data Quality

See Fig. A.25 and Table A.1.

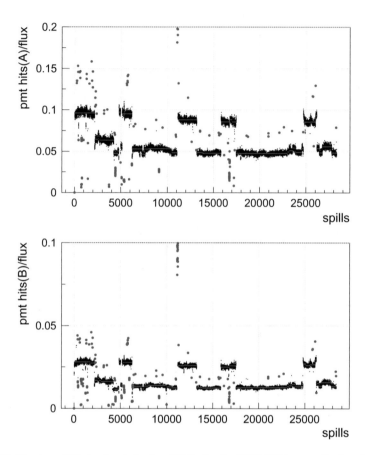

Fig. A.25 Number of hits in ring A (top) and B (bottom) of CAMERA normalised to the muon flux as a function of the spill number. The excluded spills are marked with the red dots, while the mean value of a certain number of spills is shown by the red lines. A spill is classified as "bad" in case it deviates more than 5 sigma from the mean value

Table A.1 Probabilities that a segment of CAMERA is operational, calculated according to Eq. (5.11)

Segment (A, B)	All data	μ^+ data	μ^- data
(0, 0)	0.941	0.956	0.929
(0, 1)	0.954	0.976	0.935
(1, 1)	0.973	0.987	0.961
(1, 2)	0.963	0.947	0.976
(2, 2)	0.928	0.941	0.917
(2, 3)	0.940	0.975	0.911
(3, 3)	0.976	0.987	0.967
(3, 4)	0.981	0.985	0.977
(4, 4)	0.981	0.985	0.977
(4, 5)	0.962	0.949	0.973
(5, 5)	0.962	0.949	0.973
(5, 6)	0.972	0.973	0.970
(6, 6)	0.967	0.971	0.963
(6, 7)	0.970	0.983	0.959
(7, 7)	0.972	0.980	0.965
(7, 8)	0.981	0.990	0.974
(8, 8)	0.983	0.995	0.972
(8, 9)	0.980	0.984	0.976
(9, 9)	0.979	0.984	0.976
(9, 10)	0.773	0.767	0.777
(10, 10)	0.593	0.561	0.616
(10, 11)	0.761	0.693	0.814
(11, 11)	0.963	0.975	0.952
(11, 12)	0.944	0.965	0.927
(12, 12)	0.955	0.965	0.947
(12, 13)	0.972	0.993	0.956
(13, 13)	0.973	0.994	0.956
(13, 14)	0.975	0.982	0.970
(14, 14)	0.729	0.613	0.822
(14, 15)	0.738	0.627	0.825
(15, 15)	0.980	0.996	0.967
(15, 16)	0.972	0.996	0.952
(16, 16)	0.980	0.996	0.967
(16, 17)	0.980	0.987	0.975
(17, 17)	0.972	0.987	0.961
(17, 18)	0.812	0.943	0.705
(18, 18)	0.564	0.561	0.564
(18, 19)	0.622	0.600	0.639
(19, 19)	0.976	0.989	0.965

Table A.1 (continued)

Segment (A, B)	All data	μ^+ data	μ^- data
(19, 20)	0.976	0.996	0.960
(20, 20)	0.975	0.996	0.958
(20, 21)	0.952	0.985	0.925
(21, 21)	0.952	0.985	0.926
(21, 22)	0.822	0.777	0.858
(22, 22)	0.642	0.576	0.694
(22, 23)	0.646	0.586	0.694
(23, 23)	0.721	0.607	0.811
(23, 0)	0.700	0.582	0.794

A.4 Event Selection

A.4.1 Muon and Vertex Selection

See Figs. A.26, A.27 and A.28.

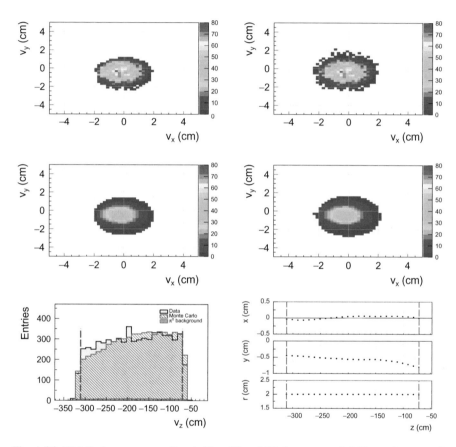

Fig. A.26 Distributions corresponding to Figs. 6.2 and 6.1 for an extended kinematic range of: $(10\,\text{GeV} < \nu < 144\,\text{GeV})$ and $(1\,(\text{GeV/c})^2 < Q^2 < 20\,(\text{GeV/c})^2)$

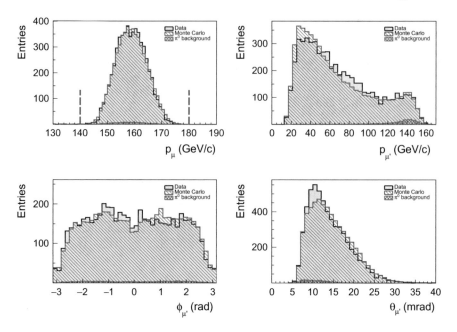

Fig. A.27 Distributions corresponding to Fig. 6.3 for an extended kinematic range of: $(10\,\mathrm{GeV} < \nu < 144\,\mathrm{GeV})$ and $(1\,(\mathrm{GeV/c})^2 < Q^2 < 20\,(\mathrm{GeV/c})^2)$

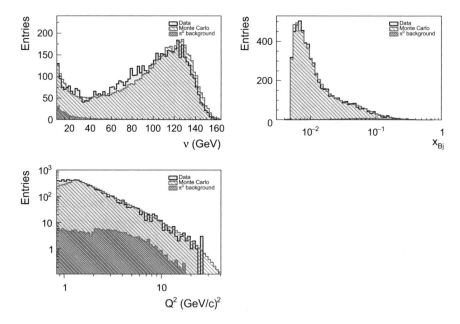

Fig. A.28 Distributions corresponding to Figs. 6.4 and 6.5 for an extended kinematic range of: $(10\,\mathrm{GeV} < \nu < 144\,\mathrm{GeV})$ and $(1\,(\mathrm{GeV/c})^2 < Q^2 < 20\,(\mathrm{GeV/c})^2)$

A.4.2   Photon Selection

See Figs. A.29 and A.30.

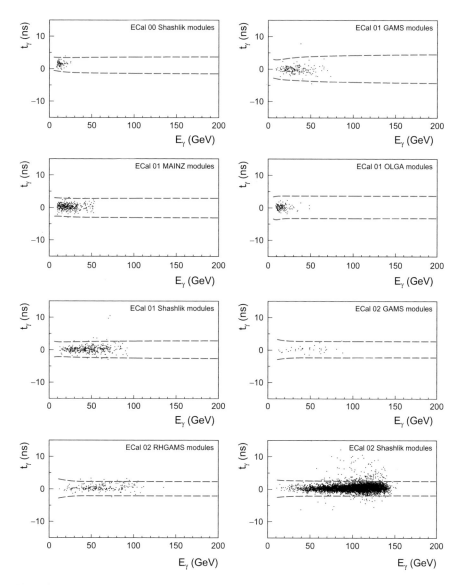

Fig. A.29 Distributions corresponding to Fig. 6.7 for an extended kinematic range of: ($10\,\mathrm{GeV} < \nu < 144\,\mathrm{GeV}$) and ($1\,(\mathrm{GeV/c})^2 < Q^2 < 20\,(\mathrm{GeV/c})^2$)

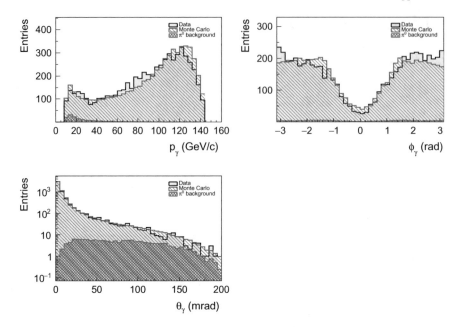

Fig. A.30 Distributions corresponding to Fig. 6.8 for an extended kinematic range of: $(10\,\mathrm{GeV} < \nu < 144\,\mathrm{GeV})$ and $(1\,(\mathrm{GeV/c})^2 < Q^2 < 20\,(\mathrm{GeV/c})^2)$

A.4.3 Proton Selection and Application of the Exclusivity Cuts

See Figs. A.31, A.32 and A.33.

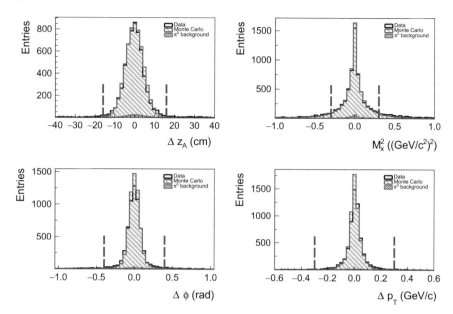

Fig. A.31 Distributions corresponding to Fig. 6.9 for an extended kinematic range of: $(10\,\mathrm{GeV} < \nu < 144\,\mathrm{GeV})$ and $(1\,(\mathrm{GeV/c})^2 < Q^2 < 20\,(\mathrm{GeV/c})^2)$

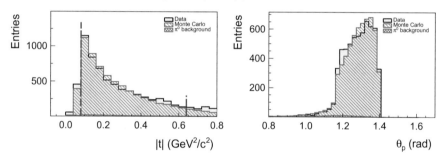

Fig. A.32 Distributions corresponding to Fig. 6.10 for an extended kinematic range of: $(10\,\mathrm{GeV} < \nu < 144\,\mathrm{GeV})$ and $(1\,(\mathrm{GeV/c})^2 < Q^2 < 20\,(\mathrm{GeV/c})^2)$

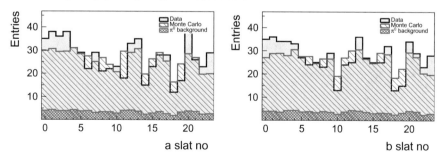

Fig. A.33 Distributions corresponding to Fig. 6.10 for the kinematic range, used for the extraction of the DVCS cross section: $(10\,\mathrm{GeV} < \nu < 32\,\mathrm{GeV})$ and $(1\,(\mathrm{GeV/c})^2 < Q^2 < 5\,(\mathrm{GeV/c})^2)$

A.4.4   The Kinematic Fit for DVCS

See Figs. A.34, A.35, A.36, A.37, A.38, A.39, A.40, A.41, A.42, A.43, A.44 and A.45.

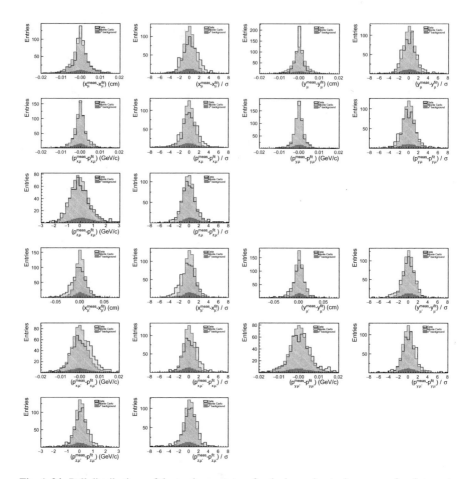

Fig. A.34 Pull distributions of the track parameters for the in- and outgoing muon after the event selection of Chap. 6: See Fig. 6.13 for the used abbreviations. The energy and momentum constraints have been shifted according to Eq. (6.15). The used kinematic range is: $(10\,\text{GeV} < \nu < 32\,\text{GeV})$ and $(1\,(\text{GeV/c})^2 < Q^2 < 5\,(\text{GeV/c})^2)$

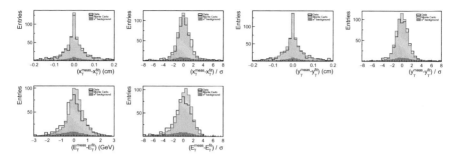

Fig. A.35 Pull distributions of the track parameters of the photon after the event selection of Chap. 6: See Fig. 6.14 for the used abbreviations. The energy and momentum constraints have been shifted according to Eq. (6.15). The used kinematic range is: $(10\,\text{GeV} < \nu < 32\,\text{GeV})$ and $(1\,(\text{GeV/c})^2 < Q^2 < 5\,(\text{GeV/c})^2)$

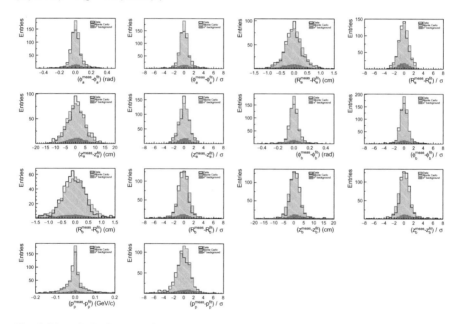

Fig. A.36 Pull distributions of the proton track parameters after the event selection of Chap. 6: See Fig. 6.15 for the used abbreviations. The energy and momentum constraints have been shifted according to Eq. (6.15). The used kinematic range is: $(10\,\text{GeV} < \nu < 32\,\text{GeV})$ and $(1\,(\text{GeV/c})^2 < Q^2 < 5\,(\text{GeV/c})^2)$

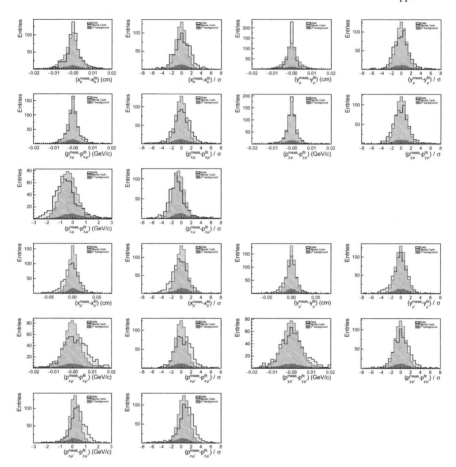

Fig. A.37 Pull distributions of the track parameters for the in- and outgoing muon after the event selection of Chap. 6: See Fig. 6.13 for the used abbreviations. No shift for the energy and momentum constraints has been applied. The used kinematic range is: $(10\,\text{GeV} < \nu < 32\,\text{GeV})$ and $(1\,(\text{GeV/c})^2 < Q^2 < 5\,(\text{GeV/c})^2)$

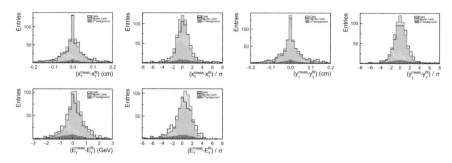

Fig. A.38 Pull distributions of the track parameters of the photon after the event selection of Chap. 6: See Fig. 6.14 for the used abbreviations. No shift for the energy and momentum constraints has been applied. The used kinematic range is: $(10\,\mathrm{GeV} < \nu < 32\,\mathrm{GeV})$ and $(1\,(\mathrm{GeV/c})^2 < Q^2 < 5\,(\mathrm{GeV/c})^2)$

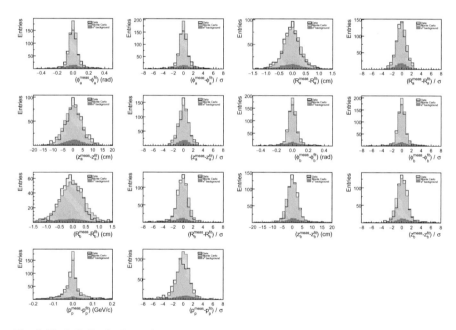

Fig. A.39 Pull distributions of the proton track parameters after the event selection of Chap. 6: See Fig. 6.15 for the used abbreviations. No shift for the energy and momentum constraints has been applied. The used kinematic range is: $(10\,\mathrm{GeV} < \nu < 32\,\mathrm{GeV})$ and $(1\,(\mathrm{GeV/c})^2 < Q^2 < 5\,(\mathrm{GeV/c})^2)$

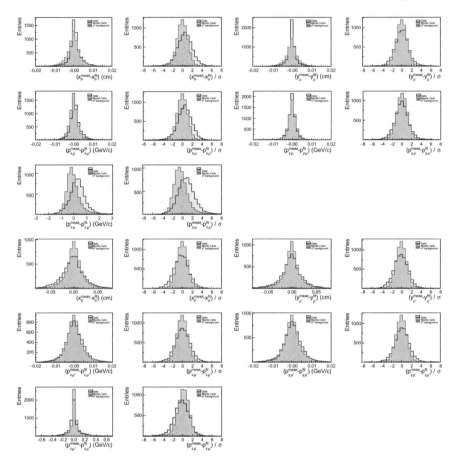

Fig. A.40 Pull distributions of the track parameters for the in- and outgoing muon after the event selection of Chap. 6: See Fig. 6.13 for the used abbreviations. The energy and momentum constraints have been shifted according to Eq. (6.15). The used kinematic range is: $(10\,\text{GeV} < \nu < 144\,\text{GeV})$ and $(1\,(\text{GeV}/c)^2 < Q^2 < 20\,(\text{GeV}/c)^2)$

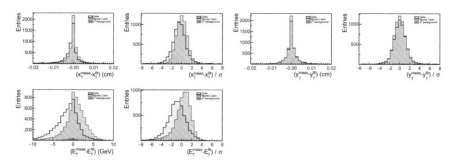

Fig. A.41 Pull distributions of the track parameters of the photon after the event selection of Chap. 6: See Fig. 6.14 for the used abbreviations. The energy and momentum constraints have been shifted according to Eq. (6.15). The used kinematic range is: $(10\,\mathrm{GeV} < \nu < 144\,\mathrm{GeV})$ and $(1\,(\mathrm{GeV/c})^2 < Q^2 < 20\,(\mathrm{GeV/c})^2)$

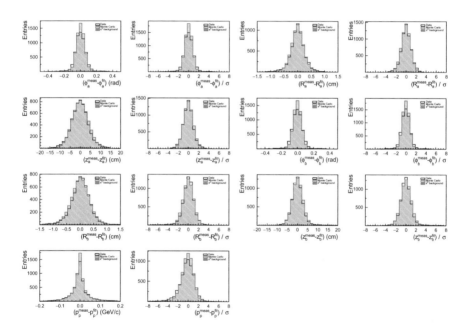

Fig. A.42 Pull distributions of the proton track parameters after the event selection of Chap. 6: See Fig. 6.15 for the used abbreviations. The energy and momentum constraints have been shifted according to Eq. (6.15). The used kinematic range is: $(10\,\mathrm{GeV} < \nu < 144\,\mathrm{GeV})$ and $(1\,(\mathrm{GeV/c})^2 < Q^2 < 20\,(\mathrm{GeV/c})^2)$

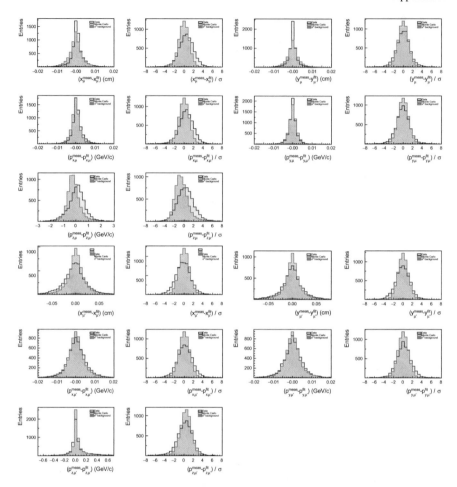

Fig. A.43 Pull distributions of the track parameters for the in- and outgoing muon after the event selection of Chap. 6: See Fig. 6.13 for the used abbreviations. No shift for the energy and momentum constraints has been applied. The used kinematic range is: $(10\,\mathrm{GeV} < \nu < 144\,\mathrm{GeV})$ and $(1\,(\mathrm{GeV/c})^2 < Q^2 < 20\,(\mathrm{GeV/c})^2)$

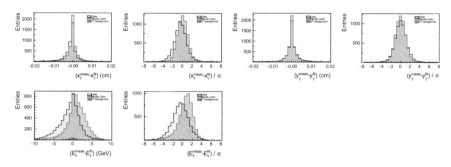

Fig. A.44 Pull distributions of the track parameters of the photon after the event selection of Chap. 6: See Fig. 6.14 for the used abbreviations. No shift for the energy and momentum constraints has been applied. The used kinematic range is: $(10\,\text{GeV} < \nu < 144\,\text{GeV})$ and $(1\,(\text{GeV/c})^2 < Q^2 < 20\,(\text{GeV/c})^2)$

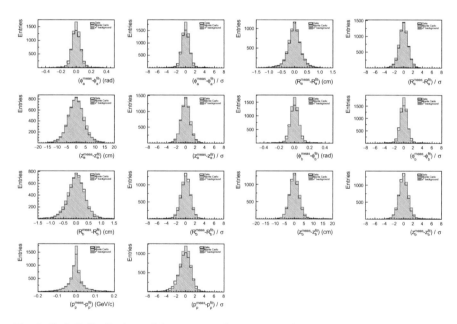

Fig. A.45 Pull distributions of the proton track parameters after the event selection of Chap. 6: See Fig. 6.15 for the used abbreviations. No shift for the energy and momentum constraints has been applied. The used kinematic range is: $(10\,\text{GeV} < \nu < 144\,\text{GeV})$ and $(1\,(\text{GeV/c})^2 < Q^2 < 20\,(\text{GeV/c})^2)$

A.5   The Cross Section and Its t-Dependence

A.5.1   Normalisation of the LEPTO and HEPGen++ π^0 Monte Carlos

See Figs. A.46, A.47, A.48, A.49, A.50 and A.51.

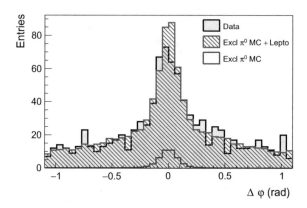

Fig. A.46   Distribution of $\Delta\phi$ for **Method 1** of Sect. 7.2.2. The blue histogram describes the overall Monte Carlo estimate given by the exclusive π^0 (HEPGen++) and the LEPTO Monte Carlo yields, while the red histogram displays the fraction described by the exclusive π^0 Monte Carlo yield

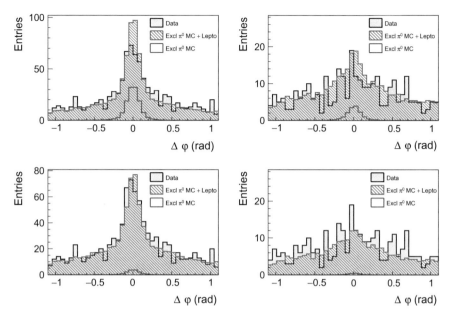

Fig. A.47 Distribution of $\Delta\phi$ for **Method 2** (top row) and **Method 3** (bottom row) of Sect. 7.2.2 for $N_B < 3$. The blue histogram describes the overall Monte Carlo estimate given by the exclusive π^0 (HEPGen++) and the LEPTO Monte Carlo yields, while the red histogram displays the fraction described by the exclusive π^0 Monte Carlo yield. Left: Set of signal distributions S. Right: Set of background like distributions B

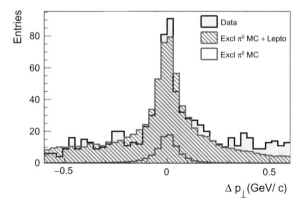

Fig. A.48 Distribution of Δp_T for **Method 1** of Sect. 7.2.2. The blue histogram describes the overall Monte Carlo estimate given by the exclusive π^0 (HEPGen++) and the LEPTO Monte Carlo yields, while the red histogram displays the fraction described by the exclusive π^0 Monte Carlo yield

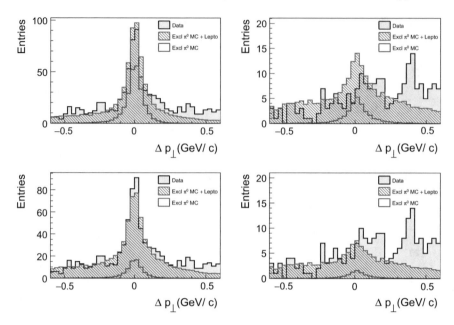

Fig. A.49 Distribution of Δp_T for **Method 2** (top row) and **Method 3** (bottom row) of Sect. 7.2.2 for $N_B < 3$. The blue histogram describes the overall Monte Carlo estimate given by the exclusive π^0 (HEPGen++) and the LEPTO Monte Carlo yields, while the red histogram displays the fraction described by the exclusive π^0 Monte Carlo yield. Left: Set of signal distributions S. Right: Set of background like distributions B

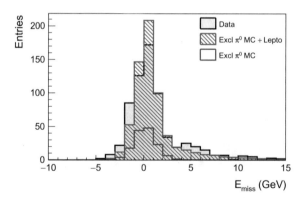

Fig. A.50 Distribution of E_{miss} for **Method 1** of Sect. 7.2.2. The blue histogram describes the overall Monte Carlo estimate given by the exclusive π^0 (HEPGen++) and the LEPTO Monte Carlo yields, while the red histogram displays the fraction described by the exclusive π^0 Monte Carlo yield

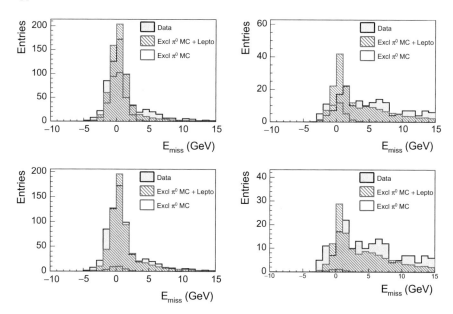

Fig. A.51 Distribution of E_{miss} for **Method 2** (top row) and **Method 3** (bottom row) of Sect. 7.2.2 for $N_B < 3$. The blue histogram describes the overall Monte Carlo estimate given by the exclusive π^0 (HEPGen++) and the LEPTO Monte Carlo yields, while the red histogram displays the fraction described by the exclusive π^0 Monte Carlo yield. Left: Set of signal distributions S. Right: Set of background like distributions B

A.5.2 The Kinematic Fit for DVCS

In order to derive Eq. (6.16) it is worth looking at the four-momentum balance: The four-momentum balance for the DVCS process, $\mu p \rightarrow \mu' p' \gamma$, reads:

$$p_\mu + p_p = p_{\mu'} + p_{p'} + p_\gamma.$$

First a small relation which comes in handy later should be derived:

$$p_\gamma(p_{\mu'} - p_\mu) = -\left(E_\gamma/c\right)\left(\nu/c - \sqrt{Q^2 + (\nu/c)^2} \cos\theta_{\gamma*\gamma}\right). \quad (A.1)$$

In order to derive this relation the four vector product is explicitly written:

$$p_\gamma(p_{\mu'} - p_\mu) = \left(E_\gamma/c, \vec{p}_\gamma\right)\left((E_{\mu'} - E_\mu)/c, (\vec{p}_\mu - \vec{p}_{\mu'})\right) = -\nu E_\gamma/c^2 + \vec{p}_\gamma(\vec{p}_\mu - \vec{p}_{\mu'}), \quad (A.2)$$

and the last part $\vec{p}_\gamma(\vec{p}_\mu - \vec{p}_{\mu'})$ is further simplified:

$$\vec{p}_\gamma(\vec{p}_\mu - \vec{p}_{\mu'}) = |\vec{p}_\gamma||\vec{p}_\mu - \vec{p}_{\mu'}| \cos\theta_{\gamma*\gamma} = E_\gamma/c\sqrt{Q^2 + (\nu/c)^2} \cos\theta_{\gamma*\gamma}, \quad (A.3)$$

while:

$$Q^2 = -(p_\mu - p_{\mu'})^2 = -(\nu/c)^2 + (\vec{p}_\mu - \vec{p}_{\mu'})^2 \Rightarrow |\vec{p}_\mu - \vec{p}_{\mu'}| = \sqrt{Q^2 + (\nu/c)^2},$$

has been used in the last step. Inserting Eq. (A.3) in (A.2) results in:

$$p_\gamma(p_{\mu'} - p_\mu) = -\nu(E_\gamma/c^2) + E_\gamma/c\sqrt{Q^2 + (\nu/c)^2} \cos\theta_{\gamma*\gamma}$$
$$= -\left(E_\gamma/c\right)\left(\nu/c - \sqrt{Q^2 + (\nu/c)^2} \cos\theta_{\gamma*\gamma}\right),$$

which coincides with Eq. (A.1). The definition of t the square of the four-momentum transfer to the proton is:

$$t := (p_p - p_{p'})^2.$$

Exploiting the four-momentum balance results in:

$$t = (p_{\mu'} - p_\mu + p_\gamma)^2 = (p_{\mu'} - p_\mu)^2 + p_\gamma^2 + 2p_\gamma(p_{\mu'} - p_\mu) = -Q^2 + 2p_\gamma(p_{\mu'} - p_\mu).$$

Inserting (A.1) for the last part, yields a formula for t, which uses the reconstructed photon energy and thus has a poor resolution:

$$t = -Q^2 - 2\left(E_\gamma/c\right)\left(\nu/c - \sqrt{Q^2 + (\nu/c)^2} \cos\theta_{\gamma*\gamma}\right). \tag{A.4}$$

In order to eliminate the quantity E_γ from the calculation of t, the four-momentum balance is written in the following way:

$$p_{p'} = p_\mu - p_{\mu'} + p_p - p_\gamma,$$

and the assumption that the recoiling target particle is a proton is exploited:

$$m_p^2 c^2 = (p_\mu - p_{\mu'} + p_p - p_\gamma)^2$$
$$= -Q^2 + (m_p c - E_\gamma/c)^2 - (E_\gamma/c)^2$$
$$+ 2\left((\nu/c)(m_p c - E_\gamma/c) + (E_\gamma/c)\sqrt{Q^2 + (\nu/c)^2} \cos\theta_{\gamma*\gamma}\right).$$

Solving for E_γ yields:

$$E_\gamma = \frac{Q^2 c^2 - 2m_p \nu c^2}{2\left(c\sqrt{Q^2 + (\nu/c)^2} \cos\theta_{\gamma*\gamma} - \nu - m_p c^2\right)}$$
$$= \frac{\nu - \frac{Q^2}{2m_p}}{1 + \frac{1}{m_p c^2}\left(\nu - c\sqrt{Q^2 + (\nu/c)^2} \cos\theta_{\gamma*\gamma}\right)}. \tag{A.5}$$

Inserting (A.5) into (A.4) results in:

$$
t = \frac{-Q^2\left(1 + \frac{1}{m_p c^2}\left(\nu - c\sqrt{Q^2 + (\nu/c)^2}\cos\theta_{\gamma*\gamma}\right)\right)}{1 + \frac{1}{m_p c^2}\left(\nu - c\sqrt{Q^2 + (\nu/c)^2}\cos\theta_{\gamma*\gamma}\right)}
$$

$$
+ \frac{\left(\frac{Q^2}{m_p c} - 2(\nu/c)\right)\left(\nu - c\sqrt{Q^2 + (\nu/c)^2}\cos\theta_{\gamma*\gamma}\right)}{1 + \frac{1}{m_p c^2}\left(\nu - c\sqrt{Q^2 + (\nu/c)^2}\cos\theta_{\gamma*\gamma}\right)}
$$

$$
= \frac{-Q^2 - 2(\nu/c)\left((\nu/c) - \sqrt{Q^2 + (\nu/c)^2}\cos\theta_{\gamma*\gamma}\right)}{1 + \frac{1}{m_p c^2}\left(\nu/c - \sqrt{Q^2 + (\nu/c)^2}\cos\theta_{\gamma*\gamma}\right)},
$$

which coincides with Eq. (6.16).

A.5.3 Cross Section Extraction Method

Starting from:

$$
\left\langle\frac{d\sigma_{DVCS}^{\gamma^* p \to \gamma p'}}{dt}\right\rangle_{ijn} = \left\langle\frac{1}{\Gamma}\frac{d\sigma_{data}^{\mu p \to \mu'\gamma p'}}{dt\,dQ^2 d\nu}\right\rangle_{ijn} - \left\langle\frac{1}{\Gamma}\frac{d\sigma_{BH}^{\mu p \to \mu'\gamma p'}}{dt\,dQ^2 d\nu}\right\rangle_{ijn} - \left\langle\frac{1}{\Gamma}\frac{d\sigma_{\pi^0}^{\mu p \to \mu'\gamma p'}}{dt\,dQ^2 d\nu}\right\rangle_{ijn},
$$

the 3 terms are given separately by:

$$
\left\langle\frac{1}{\Gamma}\frac{d\sigma_{data}^{\mu p \to \mu'\gamma p'}}{dt\,dQ^2 d\nu}\right\rangle_{ijn} \approx \left\langle\frac{1}{\Gamma}\right\rangle_{ijn}\left\langle\frac{d\sigma_{data}^{\mu p \to \mu'\gamma p'}}{dt\,dQ^2 d\nu}\right\rangle_{ijn},
$$

$$
\left\langle\frac{1}{\Gamma}\frac{d\sigma_{BH}^{\mu p \to \mu'\gamma p'}}{dt\,dQ^2 d\nu}\right\rangle_{ijn} \approx \left\langle\frac{1}{\Gamma}\right\rangle_{ijn}\left\langle\frac{d\sigma_{BH}^{\mu p \to \mu'\gamma p'}}{dt\,dQ^2 d\nu}\right\rangle_{ijn},
$$

$$
\left\langle\frac{1}{\Gamma}\frac{d\sigma_{\pi^0}^{\mu p \to \mu'\gamma p'}}{dt\,dQ^2 d\nu}\right\rangle_{ijn} \approx \left\langle\frac{1}{\Gamma}\right\rangle_{ijn}\left\langle\frac{d\sigma_{\pi^0}^{\mu p \to \mu'\gamma p'}}{dt\,dQ^2 d\nu}\right\rangle_{ijn}.
$$

The beam charge $\pm$ is omitted here for clarity. The approximation in these three equations can either be justified by assuming that the cross section is approximately constant on the bin or by Sect. A.5.4.

Transforming these equations a bit more one can see how the acceptance enters and what is technically done during the extraction procedure:

$$
\left\langle\frac{1}{\Gamma}\right\rangle_{ijn}\left\langle\frac{d\sigma_{\text{data}}^{\mu p\to\mu'\gamma p'}}{dt\,dQ^2 d\nu}\right\rangle_{ijn}=\frac{\left\langle\frac{1}{\Gamma}\right\rangle_{ijn}N_{ijn}^{\text{data}}(a_{ijn})^{-1}}{\Delta t_n\,\Delta Q_i^2\,\Delta\nu_j\mathcal{L}}=\frac{(\sum_e^{N_{ijn}^{\text{data}}}\frac{1}{\Gamma^e})(a_{ijn})^{-1}}{\Delta t_n\,\Delta Q_i^2\,\Delta\nu_j\mathcal{L}},
$$

$$
\left\langle\frac{1}{\Gamma}\right\rangle_{ijn}\left\langle\frac{d\sigma_{\text{BH}}^{\mu p\to\mu'\gamma p'}}{dt\,dQ^2 d\nu}\right\rangle_{ijn}=c_{BH}\cdot\frac{\left\langle\frac{1}{\Gamma}\right\rangle_{ijn}W_{ijn}^{\text{BH}}(a_{ijn})^{-1}}{\Delta t_n\,\Delta Q_i^2\,\Delta\nu_j\mathcal{L}}=c_{BH}\cdot\frac{(\sum_e^{N_{ijn}^{\text{BH}}}\frac{(w_{\text{P.A.M.}})_e}{\Gamma^e})(a_{ijn})^{-1}}{\Delta t_n\,\Delta Q_i^2\,\Delta\nu_j\mathcal{L}},
$$

$$
\left\langle\frac{1}{\Gamma}\right\rangle_{ijn}\left\langle\frac{d\sigma_{\pi^0}^{\mu p\to\mu'\gamma p'}}{dt\,dQ^2 d\nu}\right\rangle_{ijn}=c_{\pi_\gamma^0}\cdot\frac{\left\langle\frac{1}{\Gamma}\right\rangle_{ijn}W_{ijn}^{\pi_\gamma^0}(a_{ijn})^{-1}}{\Delta t_n\,\Delta Q_i^2\,\Delta\nu_j\mathcal{L}}=c_{\pi_\gamma^0}\cdot\frac{(\sum_e^{N_{ijn}^{\pi_\gamma^0}}\frac{(w_{\pi_\gamma^0})_e}{\Gamma^e})(a_{ijn})^{-1}}{\Delta t_n\,\Delta Q_i^2\,\Delta\nu_j\mathcal{L}},
$$

while the abbreviations:

$$
W_{ijn}^{\pi_\gamma^0}:=\sum_e^{N_{ijn}^{\pi_\gamma^0}}(w_{\pi_\gamma^0})_e\text{ and }W_{ijn}^{\text{BH}}:=\sum_e^{N_{ijn}^{\text{BH}}}(w_{\text{P.A.M.}})_e,
$$

have been used.

A.5.4 Event by Event Calculation of the Transverse Virtual Photon Flux

The following term has to be evaluated:

$$
\left\langle\frac{1}{\Gamma}\frac{d\sigma}{d\Omega}\right\rangle_{\Delta\Omega}.
$$

while $\Delta\Omega=(\Delta Q_i^2\Delta\nu_j)$ and σ is short-handed for $\sigma^{\mu p\to\mu'\gamma p'}$. If one subdivides the bin $\Delta\Omega$ in sub bins $\Delta\Omega_k$ one can write:

$$
\left\langle\frac{1}{\Gamma}\frac{d\sigma}{d\Omega}\right\rangle_{\Delta\Omega}=\frac{\sum_k\left\langle\frac{1}{\Gamma}\frac{d\sigma}{d\Omega}\right\rangle_{\Delta\Omega_k}}{\sum_k\Delta\Omega_k},
$$

which is simply the weighted mean over the sub bins.

If one now chooses the sub binning such that one finds exactly one or zero events in each sub bin $\Delta\Omega_k$, and if one lets the sub bins where one has observed an event being sufficiently small, one can transform the term above as follows:

$$\frac{\sum_k \left\langle \frac{1}{\Gamma} \frac{d\sigma}{d\Omega} \right\rangle_{\Delta\Omega_k}}{\sum_k \Delta\Omega_k} = \frac{\sum_e \frac{1}{\Gamma(Q_e^2, \nu_e)} \frac{1}{\Delta\Omega_e \mathcal{L}} \Delta\Omega_e}{\sum_k \Delta\Omega_k} = \frac{\sum_e \frac{1}{\Gamma(Q_e^2, \nu_e)}}{\Delta\Omega\mathcal{L}} = \frac{\sum_e \frac{1}{\Gamma(Q_e^2, \nu_e)}}{N_{\Delta\Omega}} \frac{N_{\Delta\Omega}}{\mathcal{L}\Delta\Omega}$$

$$= \left\langle \frac{1}{\Gamma} \right\rangle_{\Delta\Omega} \left\langle \frac{d\sigma}{d\Omega} \right\rangle_{\Delta\Omega}.$$

Thus, one can see that the approximation of Sect. A.5.3 can be justified in the discrete case, which is unavoidable for a binned cross section extraction.

A.5.5 Cross Section Extraction Using a Binned Calculation of the Transverse Virtual Photon Flux

In Sect. 7.1 the background and Bethe–Heitler correction was treated on the level of cross sections. In this approach the number of events in each bin is corrected for the Bethe–Heitler contribution and the π^0 contamination:

$$N_{ijn}^{\pm} = N_{ijn}^{\text{data},\pm} - N_{ijn}^{\text{BH},\pm} - N_{ijn}^{\pi^0,\pm}.$$

The DVCS cross section of the bin (i, j, n) now reads:

$$\left\langle \frac{d\sigma_{DVCS}^{\gamma^* p \to \gamma p'}}{d|t|} \right\rangle_{ijn}^{\pm} = \frac{N_{ijn}^{\pm} (a_{ijn}^{\pm})^{-1}}{\Delta t_n \Delta Q_i^2 \Delta \nu_j \mathcal{L}^{\pm}} \left(\frac{1}{\Gamma_{DVCS}^{MC}(\hat{Q}^2, \hat{\nu})} \right).$$

This is summed according to Eqs. (7.1) and (6.6). The factor $\left(\frac{1}{\Gamma_{DVCS}^{MC}(\hat{Q}^2, \hat{\nu})} \right)$, which is the virtual-photon flux evaluated at the mean Q^2 and mean ν of each bin, using a model dependant MC for the DVCS process, is used to transform from a muon proton to a virtual-photon proton cross section. It should be emphasised that this is not the favoured procedure due to its model dependence, but should rather be seen as a consistency check of the procedure, described in Sect. 7.1.

A.5.6 The DVCS Cross Section and the Extraction of the t-Slope

See Table A.2.

Table A.2 Values of the extracted DVCS cross section and mean kinematic quantities: The quantity $\frac{d\sigma}{d|t|}$ denotes the mean differential DVSCS cross section in the indicated $|t|$-bin. The statistical uncertainty is denoted by S, while the systematic uncertainties are denoted by $S^\uparrow$ and $S^\downarrow$. The arrow indicates the direction of the systematic uncertainties

$|t|$-bin/(Gev/c)^2	]0.08, 0.22]	]0.22, 0.36]	]0.36, 0.5]	]0.5, 0.64[	]0.08, 0.64[
$\frac{d\sigma}{d|t|}$ /nb(GeV/c)$^{-2}$	24.54	12.58	7.40	4.05	12.14
$S^\uparrow$/nb(GeV/c)$^{-2}$	3.73	2.24	1.29	0.95	1.16
$S^\downarrow$/nb(GeV/c)$^{-2}$	2.89	1.45	0.85	0.48	0.84
S/nb(GeV/c)$^{-2}$	2.82	1.98	1.55	1.32	1.00
$< W >$/GeV/c^2	5.89	5.79	5.70	5.99	5.84
$< Q^2 >$/(Gev/c)^2	1.79	1.77	1.91	1.77	1.80
$< x_{Bj} >$	0.054	0.055	0.065	0.055	0.056
$< \nu >$/GeV	19.48	18.82	18.56	20.14	19.22
$< \xi >$	0.028	0.028	0.0634	0.029	0.029

A.5.6.1 Toy Monte Carlo Check for the t-Slope Estimator

The purpose of this section is to check if the binned maximum likelihood fit gives a good estimator for the t-slope and if the statistical error given on the t-slope is at a reasonable scale. In the signal region one detects in total 649 events. From the Monte Carlo one estimates that 278 events are due to the Bethe–Heitler process and the π^0 background. Thus, one has 371 events left, which are to be considered as signal. If one assumes that the Monte Carlo statistics is sufficiently large such that the background correction does not introduce further statistical fluctuations, one is left with a relative statistical error on the number of signal events as follows:

$$S_r = \frac{\sqrt{649}}{371},$$

which would correspond to measuring

$$N = \frac{1}{S_r^2} = 212$$

events if one assumes to have no background. Thus, in the following a toy Monte Carlo study is presented for which 212 random exponentially distributed events with a t-slope value of 4.3 are generated 10000 times. Each sample is fitted with a χ^2 fit, a maximum likelihood fit included in ROOT and the binned Maximum Likelihood procedure described in Sect. 7.5.

Looking at Fig. A.52 one observes that the χ^2 fit is biased and tends to have larger statistical errors on the result for the slope. Furthermore, one observes that both maximum likelihood fits give valid estimators for the slope parameter.

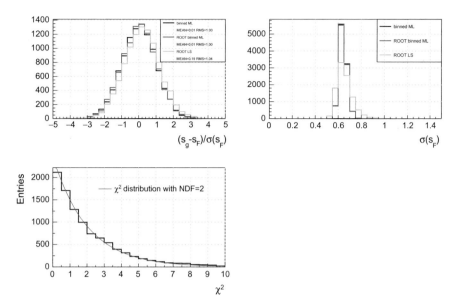

Fig. A.52 The quantity s_g is the generated value of the slope, s_f the estimator for the slope, $\sigma(S_F)$ the estimated error on the slope, given by the different procedures. Upper left: Pull distribution for the different fitting procedures; Upper right: Distribution of $\sigma(S_F)$ for the different procedures Lower left: The χ^2 distribution of the binned maximum likelihood procedure described in Sect. 7.5

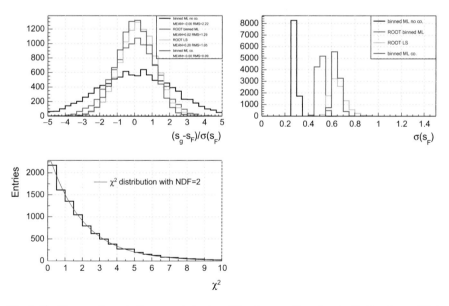

Fig. A.53 The quantity s_g is the generated value of the slope, s_f the estimator for the slope, $\sigma(S_F)$ the estimated error on the slope, given by the different procedures. For these distributions each event was scaled by a factor of 5. Upper left: Pull distribution for the different fitting procedures; Upper right: Distribution of $\sigma(S_F)$ for the different procedures Lower left: The χ^2 distribution of the binned maximum likelihood procedure described in Sect. 7.5

For Fig. A.53 each event was scaled by a factor of 5. The purpose of this exercise is to see if the different estimators given for the slope are still valid if one scales the events as it has to be done during the extraction procedure of the t-slope. One observes that the behaviour of the estimator given by the χ^2 fit does not change. The maximum likelihood estimator given by ROOT seems to produce errors which tend to be too small, which can be seen by looking at the RMS of the red pull distribution. Furthermore, the maximum likelihood estimator constructed as described in Sect. 7.5 without the correction of the statistical error gives completely unreasonable uncertainties. However, after applying the error correction it becomes a valid estimator.

Figures A.52 and A.53 show the corresponding χ^2 values which nicely follow a χ^2 distribution with two degrees of freedom. Thus, one can conclude that the χ^2 is constructed correctly in both cases.

Finally, looking at the statistical errors shown in Figs. A.52 and A.53 one can see that one would estimate a statistical error at the order of 0.6–0.7 for the t-slope. This is in reasonable agreement with the statistical error given in Sect. 7.5, which is in case of the kinematic fit 0.62.

A.5.6.2   Statistical Fluctuations for the Extracted DVCS Cross Section

See Figs. A.54, A.55 and A.56.

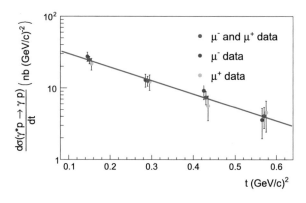

Fig. A.54 Figure 7.10 separated for the two beam charges

Fig. A.55 Figure 7.10 in the range $(10\,\mathrm{GeV} < \nu < 20\,\mathrm{GeV})$

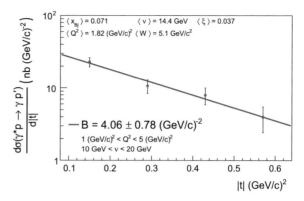

Fig. A.56 Figure 7.10 in the range $(20\,\mathrm{GeV} < \nu < 32\,\mathrm{GeV})$

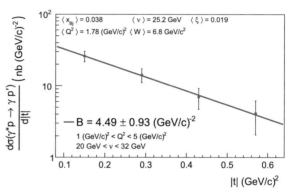

A.5.7 Impact of a Binned Calculation of the Transverse Virtual Photon Flux

This section does not contribute to the systematic error. It has the purpose to gain confidence in the extraction method. Figures A.57 and A.58 show the influence on the extracted values when one uses the alternative extraction method of Sect. A.5.5 for which the virtual photon flux is not calculated event by event from the data but taken from the Monte Carlo resp. the DVCS model included in HEPGen++ in a binned fashion. This is a strong evidence that the DVCS model in HEPGen++ describes the data very reasonably and that one observes no strong influence on the way one treats the virtual photon flux for the transition from muon proton to virtual photon proton cross section.

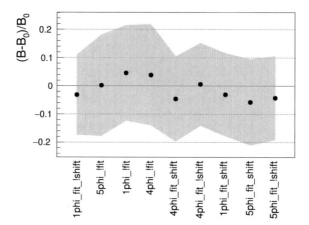

Fig. A.57 Influence on the extraction of the t-slope for different scenarios. The alternative extraction method of Sect. A.5.5 is used: nphi denotes the number of equidistant $\phi_{\gamma * \gamma}$ bins used for a 4 dimensional acceptance binning, fit/!fit denotes if the kinematic fit is used or not, shift/!shift denotes if the energy and momentum conservation of the kinematic fit is strictly zero or put to the values of Sect. 6.3. B_0 denotes the preferred value of the t-slope. The plot is normalised to this value

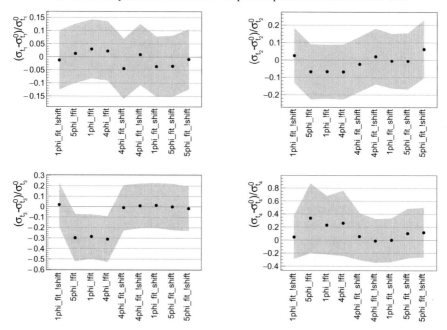

Fig. A.58 Influence on the extraction of the cross section in the four bins of t for different scenarios. The alternative extraction method of Sect. A.5.5 is used: nphi denotes the number of equidistant $\phi_{\gamma * \gamma}$ bins used for a 4 dimensional acceptance binning, fit/!fit denotes if the kinematic fit is used or not, shift/!shift denotes if the energy and momentum conservation of the kinematic fit is strictly zero or put to the values of Sect. 6.3. $\sigma_{t_i}^0$ denotes the preferred value of the extracted cross section in the corresponding t-bin with $i \in \{1, 2, 3, 4\}$. Each plot is normalised to this value

A.6   Time Synchronisation of the GANDALF Module

See Fig. A.59.

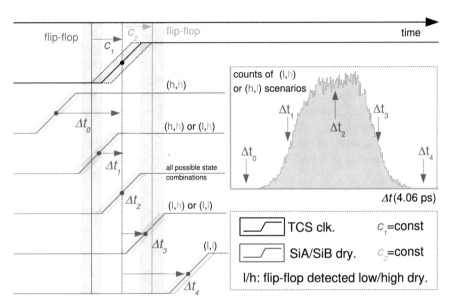

Fig. A.59 Illustration of the emergence of the grey distribution used to align the phase of the SiA/B data ready signals to the clock provided by the COMPASS trigger control system (TCS): Two flip-flops, shown in magenta and green, are driven with the TCS clock and can thus be regarded to have fixed latency with respect to the TCS clock, indicated by the constant time offsets c_1 and c_2. Three states should be distinguished: For a shift of the SiA/B data ready signal of Δt_0 or Δt_4 both flip-flops detect a (**high**, **high**) respectively (**low**, **low**) state of the SiA/B data ready signal since the rising edge of the SiA/B data ready signal is far enough away from the sampling region of the flip-flops, taking into account the jitter on the signals, shown by the transparent areas. For a shift of Δt_1 or Δt_3 one of the two flip-flops samples the edge of the SiA/B data ready signal, while the other one samples a pure high or respectively low state of the signal. Since the state of the flip-flop, sampling the rising edge, can be either low or high, a different state of the two flip-flops is observed with a certain probability. In case of a shift of Δt_2 both flip-flops are sampling the rising edge of the SiA/B data ready signal within the jitter and a plateau like behaviour is observed within the grey distribution, shown at the top right. Each of the scenarios corresponding to a shift Δt_i is recorded $\sim 6 \cdot 10^4$ times and the number of occurrences of mixed flip-flop states, either (**high**,**low**) or (**low**, **high**) is shown within the grey distribution

A.7   CAMERA Detector Commissioning 2016

In order to monitor the operation of the CAMERA detector during the data taking, a first calibration of the detector has to be achieved. Furthermore, to correctly set the high voltage of the ring A photomultipliers, a good knowledge of the amplitude spectra as a function of the longitudinal hit position within ring A is crucial.

These tasks have been achieved by combining the data given by a laser system, which simultaneously injects a light pulse in the middle of the 48 scintillators, with a measurement of cosmic muons traversing the detector. In addition, the good knowledge about the response of the outer ring of scintillators, which was not touched between 2012 and 2016, can be exploited. A detailed description of the procedure is layed down in Ref. [1]. It shall only be briefly summarised here.

To ensure a fixed reference point of the time measurements, the first step is to determine the laser reference constants c_{Ai}, c_{Bi}, k_{Ai} and k_{Bi} with the laser system:

- Laser reference constants for the time difference:

$$< t_{Ai}^u - t_{Ai}^d > + c_{Ai} = 0,$$

$$< t_{Bi}^u - t_{Bi}^d > + c_{Bi} = 0.$$

- Laser reference constants for the absolute time measurement:

$$< (t_{Ai}^u + t_{Ai}^d)/2 > + k_{Ai} = 0,$$

$$< (t_{Bi}^u + t_{Bi}^d)/2 > + k_{Bi} = 0.$$

The mean values of the respective distributions are denoted by $<>$. The index $i \in \{0, \ldots, 23\}$ indicates the scintillator number and $t_{Ai;Bi}^u$, $t_{Ai;Bi}^d$ the measured time-stamps of the photomultiplier pulses, detected at the up- or downstream side of a ring A or B scintillator. The constants c_{Ai}, c_{Bi}, k_{Ai} and k_{Bi} might change if one exchanges a photomultiplier or a signal cable between photomultiplier and digitiser. In case the internal offsets of the readout electronics like e.g. the time-stamp S, given by Eq. (9.2.2), have to be changed, the constants k_{Ai} and k_{Bi} must be reextracted from laser data.

The next step is more time consuming, but has to be performed only once. Cosmic muons, traversing subsequently a ring B(A) and a ring A(B) element perpendicular to the surface of the scintillator, are selected. Denoting the measured time-stamps in ring A and B with respect to the laser reference constants above as:

$$t_{Ai} := (t_{Ai}^u + t_{Ai}^d)/2 + k_{Ai},$$

$$t_{Bi} := (t_{Bi}^u + t_{Bi}^d)/2 + k_{Bi},$$

the offset $k_{l,m}^{ToF}$ is determined by requiring:

$$ToF := d_{AB}/c_\mu = <t_{Bl} - t_{Am}> \pm k_{l,m}^{ToF}.$$

The $\pm$ ensures the correct chronological order, as shown on the left schematic drawing of Fig. A.60. The quantity d_{AB} denotes the shortest distance between ring A and B, the quantity c_μ the speed of a cosmic muon, assumed to be the speed of light. The indices l, m satisfying:

$$l \in \{0, \dots, 23\} \text{ and } m \in \{l, (l+1) \bmod 24\} \text{ for a given } l,$$

indicate the 48 different possible combinations of ring A and B elements. It was found that the values of $k_{l,m}^{ToF}$ are at the order of 9 ns, which shows that it is not possible to perform the time calibration of the detector using the laser system only.

The high voltage calibration was also achieved by measuring cosmic muons. The procedure is illustrated by the right schematic drawing of Fig. A.60. Here, Δt_{Bi} and Δt_{Ai} are defined with respect to the laser reference constants:

$$\Delta t_{Ai} := t_{Ai}^u - t_{Ai}^d + c_{Ai},$$

$$\Delta t_{Bi} := t_{Bi}^u - t_{Bi}^d + c_{Bi}.$$

In this case it was required that the muon traverses two ring B elements opposite to each other. The knowledge of the longitudinal hit positions,

$$z_{Bi} = \frac{1}{2} c_{Bi} \Delta t_{Bi},$$

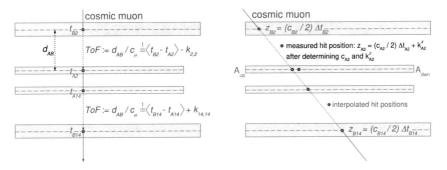

Fig. A.60 Illustration of the first order time and distance of flight calibration of the CAMERA detector in 2016. Left: The time of flight calibration is achieved by a selection of cosmic muons traversing the detector perpendicular to the surface of the scintillators. Right: The distance of flight calibration is achieved by an interpolation between the longitudinal positions z_{Bi} of ring B. The digitised amplitudes A_{up} and A_{dwn} of the photomultipliers have been studied as a function of the interpolated longitudinal hit positions in ring A in order to set the most appropriate values of the high voltage for the photomultipliers. Variables are defined according to Sect. A.7

within the ring B elements relies on the time difference calibration with the laser system and the effective speed of light c_{Bi} within ring B. By interpolating between the two longitudinal hit positions of the ring B elements the hit position inside ring A is determined. An analysis of the signal amplitudes A_{up} and A_{dwn} as a function of the interpolated longitudinal hit positions allows to extract the attenuation length of the counters. Thus, the values of the high voltage of the ring A elements can be set to the desired mean signal amplitude a minimal ionising particle would cause at a certain longitudinal position. Furthermore, analysing the measured time differences in ring A as a function of the interpolated longitudinal positions, the effective speed of light c_{Ai} and the absolute longitudinal hit position inside a ring A element with respect to ring B:

$$z_{Ai} = \frac{1}{2}c_{Ai}\Delta t_{Ai} + k_{Ai}^z,$$

can be determined. The sets of constants c_{Ai} and k_{Ai}^z are necessary to calculate the distance of flight of a particle traversing ring A and B according to Eq. (3.2).

Subsequent to this calibration procedure a pion beam was used at the COMPASS facility, in order to fine-tune the high voltage calibration of the detector. In contrast to a measurement with muon beam this allows to quickly accumulate a lot of statistics of recoiling target protons. Figure A.61 shows the energy loss in ring B as a function of β, given according to Eq. (3.4), for an exemplary ring A and ring B combination. This was recorded with the online monitoring system during the first pion run in 2016.

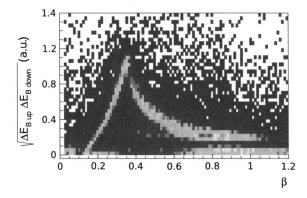

Fig. A.61 Energy loss of a proton in ring B of the CAMERA detector as a function of β, given according to Eq. (3.4). The quantities $\Delta E_{B,up}$ and $\Delta E_{B,down}$ are directly proportional to the measured signal amplitudes at the up- and downstream side of ring B. They are scaled arbitrarily within this figure. The data has been recorded, using a pion beam centred on a liquid hydrogen target surrounded by the two rings of scintillators of the recoil detector CAMERA. The rising edge of the signal describes protons being stopped in ring B, while the falling edge corresponds to protons traversing ring B and leaving the detector

Reference

1. S. Scherrers, The CAMERA Detector for the COMPASS-II Experiment at CERN - Time and Amplitude Calibrations with Lasers and Muons, Bachelor thesis, Albert Ludwigs Universität Freiburg (2016)

Printed in the United States
By Bookmasters